SOLUTIONS MANUAL

TO ACCOMPANY

BECKER'S WORLD OF THE CELL

SOLUTIONS MANUAL

TO ACCOMPANY

BECKER'S WORLD OF THE CELL

EIGHTH EDITION

Jeff Hardin
University of Wisconsin–Madison

Gregory Bertoni
The Plant Cell

Lewis J. Kleinsmith
University of Michigan, Ann Arbor

Benjamin Cummings

Boston Columbus Indianapolis New York San Francisco Upper Saddle River
Amsterdam Cape Town Dubai London Madrid Milan Munich Paris Montréal Toronto
Delhi Mexico City São Paulo Sydney Hong Kong Seoul Singapore Taipei Tokyo

Vice President/Editor-in-Chief: Beth Wilbur
Acquisitions Editor: Josh Frost
Project Editor, Biology: Anna Amato
Executive Marketing Manager: Lauren Harp
Managing Editor, Production: Michael Early
Production Project Manager: Jane Brundage
Manufacturing Buyer: Michael Penne
Production Management: Linda Kern, Progressive Publishing Alternatives
Compositor: Progressive Information Technologies
Cover Production: Seventeenth Street Studios
Text and Cover Printer: Edwards Brothers

Cover Image: Cleopatra Flesh (1962) by Jules Olitski (1922–2007); Digital Image © The Museum of Modern Art/Licensed by SCALA / Art Resource, NY.

ISBN-13: 978-0-321-68961-0
ISBN-10: 0-321-68961-5

Benjamin Cummings
is an imprint of

www.pearsonhighered.com

2 3 4 5 6 7 8 9 10 --EBM-- 14 13

Contents

CHAPTER **1**   A Preview of the Cell   1

CHAPTER **2**   The Chemistry of the Cell   5

CHAPTER **3**   The Macromolecules of the Cell   8

CHAPTER **4**   Cells and Organelles   14

CHAPTER **5**   Bioenergetics: The Flow of Energy in the Cell   18

CHAPTER **6**   Enzymes: The Catalysts of Life   25

CHAPTER **7**   Membranes: Their Structure, Function, and Chemistry   34

CHAPTER **8**   Transport Across Membranes: Overcoming the Permeability Barrier   38

CHAPTER **9**   Chemotrophic Energy Metabolism: Glycolysis and Fermentation   44

CHAPTER **10** Chemotrophic Energy Metabolism: Aerobic Respiration   52

CHAPTER **11** Phototropic Energy Metabolism: Photosynthesis   60

CHAPTER **12** The Endomembrane System and Peroxisomes   64

CHAPTER **13** Signal Transduction Mechanisms: I. Electrical Signals in Nerve Cells   69

CHAPTER **14** Signal Transduction Mechanisms: II. Messengers and Receptors   72

CHAPTER **15** Cytoskeletal Systems   74

CHAPTER **16** Cellular Movement: Motility and Contractility   76

CHAPTER **17** Beyond the Cell: Extracellular Structures, Cell Adhesion, and Cell Junctions   79

CHAPTER **18** The Structural Basis of Cellular Information: DNA, Chromosomes, and
                the Nucleus   83

CHAPTER **19** The Cell Cycle, DNA Replication, and Mitosis   87

CHAPTER **20** Sexual Reproduction, Meiosis, and Genetic Recombination   92

CHAPTER **21** Gene Expression I: The Genetic Code and Transcription   95

CHAPTER **22** Gene Expression II: Protein Synthesis and Sorting   99

CHAPTER **23** The Regulation of Gene Expression   102

CHAPTER **24** Cancer Cells   106

Preface

This *Solutions Manual* provides solutions for all of the problems that appear at the end of each chapter in the Eighth Edition of *The World of the Cell*. The inclusion of problem sets in the text reflects our conviction that one learns science not just by reading or hearing about it but by working with it. These solutions are meant to confirm your understanding when you arrive at the correct answer and to provide guidance in the appropriate problem-solving skills when you do not.

As co-authors, we retain responsibility for any errors of omission or commission, and we welcome feedback from users of this manual, students and instructors alike. Please address comments and suggestions to Jeff Hardin at jdhardin@wisc.edu, and he will direct them to the appropriate co-author.

Jeff Hardin,
Gregory Bertoni,
Lewis Kleinsmith

CHAPTER

1

A Preview of the Cell

1-1. (a) C   (d) B   (g) G   (j) G
     (b) B   (e) G   (h) B, C   (k) C
     (c) G   (f) C   (i) B   (l) B

1-2. (a) Bacterial cell: $V = \pi r^2 h = (3.14)(0.5)^2(2.0) = $ **1.57 μm^3**.

   Liver cell: $V = 4\pi r^3/3 = 4(3.14)(10)^3/3 = $ **4200 μm^3**.

   Palisade cell: $V = \pi r^2 h = (3.14)(10)^2(35) = $ **11,000 μm^3**.

(b) Bacterial cells in a liver cell: $4200/1.57 = $ **2700**.

(c) Liver cells in a palisade cell: $11,000/4200 = $ **2.62**.

1-3. (a) Light microscope: limit of resolution = 200 nm. Because 1 membrane has a thickness of about 8 nm, the number of membranes that must be aligned laterally is $200/8 = $ **25 membranes.**

   Electron microscope: limit of resolution = 0.1–0.2 nm. With a thickness of about 8 nm, a single membrane can be seen using an electron microscope.

b) Liver cell: $V = 4200$ μm^3 (from Problem 1-2a).

   Ribosome: $V = 4\pi r^3/3 = 4(3.14)(0.0125)^3/3 = 8.2 \times 10^{-6}$ μm^3.

   Ribosomes in a liver cell: $4200/(8.2 \times 10^{-6}) = 5 \times 10^8$.

(c) Bacterial cell: $V = 1.57$ μm^3 (from Problem 1-2a).

   DNA molecule: $V = \pi r^2 h = (3.14)(1 \text{ nm})^2(1.36 \text{ mm})$

   $= (3.14)(0.001$ μm$)^2(1.36 \times 10^3$ μm$)$

   $= (3.14)(1 \times 10^{-6}$ μm$^2)(1.36 \times 10^3$ μm$)$

   $= (3.14)(1.36)(10^{-3})$ μm$^3 = 4.3 \times 10^{-3}$ μm^3.

   DNA $= (4.3 \times 10^{-3})/1.57 = 0.0027 = $ **0.27% of cell volume.**

1-4. (a) The *limit of resolution* of a microscope is a measure of how close together two points can be and still be distinguished from one another when viewed through the microscope. (Note that the limit of resolution is related inversely to magnification; the greater the magnification of a microscope, the smaller its limit of resolution will be.) Hooke's microscope could only magnify objects 30-fold, so its limit of resolution was one-thirtieth that of the human eye: 0.25 mm/30 = 0.0083 mm, or 8.3 μm. Van Leeuwenhoek's microscope could magnify objects 300-fold, which was ten times greater magnification than that of Hooke's microscope, so the limit of resolution of van Leeuwenhoek's microscope was one-tenth that of Hooke's microscope: 8.3 μm/10 = 0.83 μm.

(b) The smallest structures that Hooke was able to see were about 8.3 mm in one dimension, which would have allowed him to see the plant and animal cells shown in Figure 1A-1 of the textbook but not a typical bacterial cell.

(c) The smallest structures that van Leeuwenhoek was able to see were about 0.83 μm in one dimension, which would have allowed him to see all three of the structures shown in Figure 1A-1.

(d) The limit of resolution of a modern light microscope is about 200–350 nm (0.2–0.35 μm) in one dimension, so structures must be at least this large in two dimensions to be visualized.

(e) The seven structures could be visualized by the indicated microscopes (H = Hooke's microscope; V = van Leeuwenhoek's microscope; C = contemporary light microscope):

Mycoplasma (d = 0.3 μm): C	Mitochondrion (1 μm × 2 μm): V, C
Liver cell (d = 20 μm): H, V, C	Peroxisome (d = 0.5 μm): C
Nucleus (d = 6 μm): V, C	Microtubule (25 nm × 1 μm): possibly C; most likely none
Chloroplast (2 μm × 8 μm): V, C	

1-5. (a) Cytological strand; electron microscopy is capable of much higher magnification and hence much greater resolution than light microscopy, thereby making it possible to visualize much smaller subcellular and even molecular structures.

(b) Biochemical strand; ultracentrifugation is capable of much higher speeds and hence much greater centrifugal force, thereby making it possible to resolve small subcellular structures and large molecules that cannot be separated by lower-speed centrifugation techniques.

(c) Genetic strand; DNA sequencing makes it possible to determine the order of individual nucleotides along a DNA strand, whereas nucleic hybridization can assess only whether there is enough sequence complementarity to allow two strands to adhere to one another under specified conditions.

(d) Genetic strand; sequencing a genome simply provides a large amount of data about the DNA present in the genome, whereas bioinformatics uses computer analysis to aid in understanding and interpreting those data in terms of gene numbers and expression.

(e)   Cytological strand; scanning electron microscopy makes it possible to visualize subcellular structures and macromolecules with a sense of depth—that is, as if in three dimensions.

(f)   Biochemical strand; electrophoresis separates molecules based on charge differences, thereby making it possible to separate molecules that are so similar in size, shape, and density that they may not be readily resolved by most chromatographic techniques.

1-6.   (a)   Initially thought to be true because animal cells do not have cell walls, which made it hard to distinguish individual cells using the crude microscopes available to early investigators; shown by Schwann (1839) to be incorrect for cartilage cells, which have well-defined boundaries of collagen fibers, and later extended to all animal cells.

        (b)   Initially thought to be true because living organisms seem to increase in complexity spontaneously, unlike other systems known to early chemists or physicists; misconception laid to rest by Wöhler's demonstration (1828) that urea, a compound made by living organisms, could be synthesized in the laboratory from an inorganic starting compound.

        (c)   Originally thought to be true because the order of nucleotide monomers in DNA was erroneously considered to be an invariant tetranucleotide repeating sequence; disproved by Avery et al. for bacteria (1944) and by Hershey and Chase for bacterial viruses (1952).

        (d)   Initially thought to be true because of the demonstration by Pasteur that yeast cells were needed for alcoholic fermentation; Buchner and Buchner showed later (1897) that extracts from yeast cells could substitute for intact cells, an effect we now know to be due to the presence in the extracts of the enzymes that catalyze the various reactions in the fermentation process.

1-7.   (a)   Consistent with early electron microscopy in which membranes appeared as two parallel electron-dense lines (thought to be the outer protein layer) separated by an unstained space (thought to be the lipid interior of the membrane); disproved by the demonstration that membrane proteins are globular structures located within the membrane, not just on its surface.

        (b)   Thought to be universally true until the discovery of Z-DNA, in which the two strands form a left-handed helix.

        (c)   Believed to be true until the discovery of RNA viruses (retroviruses). Retrovirus-infected cells contain virally encoded reverse transcriptase, which uses an RNA template to synthesize DNA. Thus, their name contains the root "retro," meaning backwards.

1-8.   The experimental process should begin by testing the hypothesis that the heartburn is due to the pizza. Rather than simply "keeping track of your eating habits for a few weeks," which is observational biology, you need to design an experimental protocol whereby you can test the hypothesis by deliberately consuming pizza on certain nights and refraining from pizza on others. (Ideally, such a protocol should ensure that the subject is not aware of which evenings pizza is consumed, to avoid the so-called "placebo effect" in which heartburn might be induced just by *knowing*—or *thinking*—that you've eaten pizza. But given the nature of pizza, it might be difficult to build this control into the experiment!) If you find a good correlation between pizza consumption and heartburn and are able to confirm this repeatedly—note the requirement for repeatable observations—it seems reasonable to conclude that your heartburn is, in fact, caused by eating pizza. Having tentatively confirmed that hypothesis, you are now ready to test the further hypothesis that the heartburn is caused specifically by one or more of the suspected ingredients. To do so, you need to design an experimental protocol whereby each ingredient is tested systematically, both alone and in combination with one or both of the other ingredients. In addition, you will need a "control," consisting of plain pizza with none of the suspected ingredients. (A total of seven possible "treatments" will be involved in your experimental design. Can you specify the seven?) Again, repeatability of the observations is an important criterion before you can conclude that one, two, or all three ingredients cause heartburn, either individually or in specific combinations (or, alternatively, that none of the ingredients appears to be causally involved). This means you will be eating a lot of pizza!

1-9.   For choosing a model organism to benefit your research, it is important to consider how closely related the model organism is to your organism, the particular processes you want to study, whether these process occur in a particular model organism, and the "hands-on" advantages of working with a certain model organism in the lab. Note that these questions do not have one particular correct answer because in research there are many unpredictable variables, and often an open mind and creativity are the keys to success. You will likely use a variety of techniques involving recombinant DNA technology, bioinformatics, genetic analysis, and biochemical purifications.

(a)   Since the oil is being produced in your algal cells, you might try to use *Chlamydomonas,* an algal model organism that is easy to grow in the lab. You could test whether *Chlamydomonas* produces a similar compound and try to isolate the genes and enzymes necessary to produce it. Or you could perhaps mutagenize ("knock out") genes in your algal cell and see which ones are needed to produce the biofuel.

(b)   You can introduce genes encoding one of your enzymes of interest into *Chlamydomonas* and perhaps cause it to produce the biofuel. Because these algae are easy to grow in large amounts in the lab, this could be a good source of a particular enzyme. You could try using *E. coli* because it is also easy to grow but, because it is a bacterium and is less closely related, it may be more difficult.

(c)   If you can get *E. coli* or *Chlamydomonas* to produce the biofuel, you may be able to isolate multienzyme complexes from these cells, since they are easy to grow and manipulate in the lab. Or, if you can isolate the algal genes producing your enzymes of interest, you can put them into *Saccharomyces,* a yeast commonly used to study protein-protein interactions. Then you might be able to test pairs of enzymes to see if they interact with each other.

(d)   To study organelles, you should use a eukaryotic organism, because bacteria such as *E. coli* do not have organelles. This may do well in yeast because it has eukaryotic secretion processes that could be helpful in isolating the oil. Or you could try *Arabidopsis,* which, because it is a plant, is more closely related to your algae than either *E. coli* or yeast.

CHAPTER

2

The Chemistry of the Cell

2-1. (a) Its valence of 4 allows a carbon atom to form multiple covalent bonds, both with other carbon atoms and also with other atoms, most notably oxygen, hydrogen, nitrogen, and sulfur, thereby generating a great diversity of molecules with a wide variety of properties.

(b) The high bond energy of the carbon-carbon bond ensures the stability of molecules that contain two or more carbon atoms. Specifically, the bond energy of carbon-carbon bonds is well above that of the most energetic portion of visible light (measured in kilocalories/einstein; see Figure 2-3 on p. 20 of the textbook), thereby ensuring that carbon-carbon (and other covalent bonds in biological molecules) cannot be broken by exposure to visible light. (Note, however, that this is not true of ultraviolet radiation because any radiation with a wavelength of about 344 nm or less has enough energy content to break carbon-carbon bonds.)

(c) The ability of a carbon atom to bond to two or more carbon atoms makes possible the generation of long chains of carbon atoms as well as ring structures, which are, of course, essential features of many biological molecules.

(d) The ability of carbon atoms to bond to hydrogen, nitrogen, and sulfur atoms not only increases the diversity of carbon-containing molecules but also accounts for the functional groups (such as methyl, amino, and sulfhydryl groups) that differ from one another in properties such as charge or asymmetric electron distribution, thereby enabling such molecules to play quite different roles in cellular reactions.

(e) The presence of asymmetric carbon atoms accounts for further structural diversity in the form of stereoisomers, which often differ from one another not only in their structural configuration but also in their biochemical properties.

2-2. (a) T; living organisms are essentially aqueous solutions containing many kinds of molecules, most of which are polar and hence readily soluble in water.

(b) T; oxygen is the ultimate electron acceptor in cellular respiration, with water as the product.

(c) F; the density of ice is less than that of water, thereby ensuring that ice will float on the surface of a body of water, where it will melt readily if the temperature of the surrounding air rises above the freezing point.

(d) T; this property explains the high specific heat and high heat of vaporization and hence the capacity of water to "buffer" cells and organisms against temperature changes.

(e)   T; this property of water allows light to penetrate readily, such that submerged photosynthetic organisms (or parts of organisms) can receive sunlight.

(f)   X; the lack of odor or taste is probably not a strategic advantage to most organisms.

(g)   T; high specific heat means that much heat is required to increase the temperature, which effectively "buffers" cells and organisms against temperature changes in response to changes in temperature of the environment.

(h)   T; high heat of vaporization means that much heat is required to convert water from a liquid to a gas, which means that organisms can be effectively cooled by evaporation of perspiration or other forms of liquid water from the skin or other surface of the organism.

2-3.   (a)   Oxygen, nitrogen, and carbon are the elements that most readily form strong multiple bonds. Hydrogen can only form one single bond, never multiple bonds.

(b)   Water has a higher specific heat than most other liquids because of extensive hydrogen bonding between water molecules. During heating, much of the absorbed energy is used to disrupt hydrogen bonds instead of to raise the temperature.

(c)   Hydrophobic oil droplets in water coalesce not because of an intrinsic attraction of oil molecules for each other but because they have no affinity for polar molecules and therefore are not soluble in water.

(d)   A hydrogen bond does not involve the sharing of electrons as in a covalent bond. In a hydrogen bond, a partially positively charged hydrogen atom is attracted to a nearby partly negatively charged atom, typically an oxygen or a nitrogen.

(e)   Biological membranes are called selectively permeable because only certain molecules pass through easily. Small, nonpolar molecules pass through membranes unassisted, and large polar molecules and ions require protein transporters in order to pass through them.

2-4.   Due to its lack of polar atoms (e.g., oxygen and nitrogen) and its symmetrical structure, benzene is highly nonpolar. Therefore, only nonpolar molecules such as lipids will readily dissolve in it. Slightly more polar molecules such as fatty acids and cholesterol will have low solubility in benzene. Polar molecules such as sugars, hydrophilic amino acids, and water will be insoluble in benzene.

2-5.   Several answers are possible in theory. You could encase the drug in a lipid-soluble vesicle (liposome) that will be easily absorbed by the cell. You could attach functional groups to the drugs that have known transporters in the cell membrane that take up molecules containing these groups. You could block the polar functional groups by attaching nonpolar groups that will be removed by known cellular enzymes once inside the cell. Perhaps you have thought of a reasonable strategy that could some day be workable!

2-6.   (a)   An amphipathic molecule has one or more hydrophilic regions and one or more hydrophobic regions; such polar molecules are important membrane constituents because the interior of the membrane is hydrophobic but the milieu on either side of the membrane is aqueous and hence hydrophilic.

(b)   A lipid monolayer would have one hydrophobic side and one hydrophilic side, but both sides of the membrane must interact with an aqueous environment. This means that both sides must be hydrophilic for the membrane to be a stable structure.

(c)   Selective permeability means that some specific ions and molecules can move across a given membrane at reasonable rates whereas others cannot. This, in turn, means that the membrane has transport proteins for some ions and molecules but not for others.

(d)   Although membranes are very impermeable to ions, K^+ ions (and other specific ions as well) can move across a membrane provided that the membrane has transport proteins—ion channels, often—that are specific for the particular ion.

(e)   Short sequences of hydrophobic amino acids—20–30 amino acids per sequence, usually—are likely to be the segments of the protein that span the membrane. For more details, see the discussion of integral membrane proteins in Chapter 7.

(f)   Carbohydrate side chains are relatively hydrophilic and will therefore associate more readily with the hydrophilic regions of a membrane protein rather than with the hydrophobic regions.

2-7.   It is this asymmetry that renders water a polar molecule; most of the desirable properties of water as a solvent depend on this polarity.

2-8.   (a)   A single enzyme or set of enzymes can be used to add each successive monomeric unit.

(b)   Water is a readily available reactant in an essentially aqueous world.

2-9.   (a)   TMV virions self-assemble spontaneously without the input of energy or information, which means that all of the information necessary to direct their assembly must be already present in the RNA and/or proteins.

(b)   The strain-specific assembly of TMV in vivo is determined by the RNA, not the coat protein.

(c)   The information necessary to direct self-assembly of TMV virions appears to reside in the coat protein monomers.

(d)   The self-assembly of TMV virions is specific for TMV RNA.

(e)   The most stable configuration for TMV virions is achieved by the 3:1 ratio of nucleotides and coat protein monomers and is therefore the product formed upon self-assembly regardless of the starting ratio of nucleotides and monomers.

2-10.   You would try to determine whether the macromolecule is made of a series of monomeric subunits, like macromolecules on Earth. If so, you would try to determine the number of different kinds and the arrangement of monomers. A repetitive pattern consisting of one or two different monomers would suggest a structural macromolecule. A seemingly random pattern consisting of several different monomers would suggest an informational macromolecule.

CHAPTER

3

The Macromolecules
of the Cell

3-1. (a) 2, 3, 6   (d) 1, 3, 9

(b) 4, 7, 9, 10   (e) 4, 7, 9, 10

(c) 4, 5, 9, 10   (f) 4, 5, 9, 10

3-2.

Bond	Amino Acids	Levels of Structure
Peptide (covalent)	All	Primary
Hydrogen	All	Secondary
Hydrophobic	Leucine	Tertiary, Quaternary
Ionic	Glutamate	Tertiary, Quaternary
Disulfide (covalent)	Cysteine	Tertiary, Quaternary

3-3. (a) Alanine; phenylalanine; glutamate; methionine (the less polar member of each pair).

(b) The free sulfhydryl group is polar and ionizable; the disulfide bond is much less polar.

3-4. (a) The amino acid glutamate is hydrophilic and ionizes at cellular pH, whereas valine is hydrophobic and nonionic. Substitution of the latter for the former is likely to change the chemical nature of that part of the molecule significantly.

(b) Aspartate is another acidic amino acid and is therefore a conservative change. Others that are unlikely to have major effects are the polar but uncharged amino acids serine, threonine, tyrosine, and cysteine.

(c) Yes, if the substitutions are always of like-for-like amino acids in terms of chemical properties. These are chemically conservative changes.

3-5. (a) To pull on both ends of an α-keratin polypeptide is to pull against the hydrogen bonds that account for its helical structure. These "give" readily, allowing the polypeptide to be stretched to its full, uncoiled length, at which point you would begin to pull against the covalent peptide bonds. For fibroin, you are pulling against the covalent peptide bonds immediately. (As an analogy, you might compare pulling on opposite ends of a coiled spring versus a straight length of uncoiled wire.)

(b)   Fibroin consists mainly of the two smallest amino acids, so it has few bulky R groups and can accommodate the constraints of a pleated sheet. Keratin, on the other hand, has most of the amino acids present, and the distance between bulky R groups is maximized when these protrude from a twisted helical shape.

3-6.   (a)   Hair proteins are first treated with a sulfhydryl reducing agent to break disulfide bonds and thereby destroy much of the natural tertiary structure and shape of the hair. After being "set" in the desired shape, the hair is treated with an oxidizing agent to allow disulfide bonds to re-form, but now between different cysteine groups, as determined by the positioning imposed by the curlers. These unnatural disulfide bonds then stabilize the desired configuration.

(b)   There are two reasons for the lack of permanence: (1) Disulfide bonds occasionally break and re-form spontaneously, allowing the hair proteins to return gradually to their original, thermodynamically more favorable shape. (2) Hair continues to grow, and the new α-keratin molecules will have the natural (correct) disulfide bonds.

(c)   There is probably a genetic difference in the positioning of cysteine groups and hence in the formation of disulfide bonds.

3-7.   (a)   DR; adenine is present in both nucleic acids.

(b)   D; thymidine is unique to DNA.

(c)   N; some RNA molecules (e.g., tRNAs) have double-stranded segments.

(d)   R; mRNA, tRNA, and rRNA are all involved in cytoplasmic protein synthesis.

(e)   DR; obviously true.

(f)   N; that describes a protein!

3-8.   (a)   Although proteins, nucleic acids, and polysaccharides are all very long polymers that are synthesized by condensation of individual monomer units, lipids are not. Lipids are relatively large molecules, but they are not synthesized from a long series of monomeric units.

(b)   The amino acid proline is not found in α helices because its R group is covalently bonded to the amino nitrogen. Once a peptide bond is formed involving proline, its amino nitrogen has no available hydrogen for forming a hydrogen bond within an α-helix.

(c)   A protein can be denatured by high temperature treatment or by extremes of pH, both of which disrupt tertiary structure.

(d)   Nucleic acids are synthesized from monomers that contain a high-energy phosphodiester bond. Thus, they are already activated and do not require a carrier molecule.

(e)   α-D-glucose and β-D-glucose are isomers differing in the position of the hydroxyl group attached to carbon 1, but they are not stereoisomers (mirror images). D-glucose and L-glucose are stereoisomers.

(f)   Fatty acids are important components of the phospholipids found in all cellular membranes. However, there are many other cellular lipids that do not contain fatty acids.

(g)   Although a polypeptide's primary sequence will determine its final folded (tertiary) structure, because there are nearly limitless ways in which even a small protein can fold, it is still not possible to predict this tertiary structure from the primary sequence.

(h)   In addition to these three types of RNA, many others have been isolated in recent years, including small interfering RNA (siRNA) and micro RNA (miRNA).

3-9.   (a)   Compared to a linear molecule, a branched-chain polymer has more termini for addition or hydrolysis of glucose units per unit volume of polymer, thereby facilitating both the deposition and mobilization of glucose by providing more sites for enzymatic activity.

(b)   Every branch point will have an α(1→6) glycosidic bond that will have to be hydrolyzed. This is handled by the presence of an additional enzyme specific for the α(1→6) bond.

(c)   Endolytic cleavage breaks the molecule infernally, creating additional ends for exolytic attack and thereby allowing the mobilization of more glucose per unit time.

(d)   Cellulose molecules are rigid, linear rods that aggregate laterally into microfibrils. Branches in the molecule would generate side chains that would almost certainly make it difficult to pack the cellulose molecules into microfibrils, thereby decreasing the rigidity and strength of the microfibrils.

3-10.   See Figure S3-1 for the structures of (a) gentiobiose, (b) raffinose, and (c) a portion of a dextran chain.

(a)   Gentiobiose:

(b)   Raffinose:

(c)   Portion of a dextran:

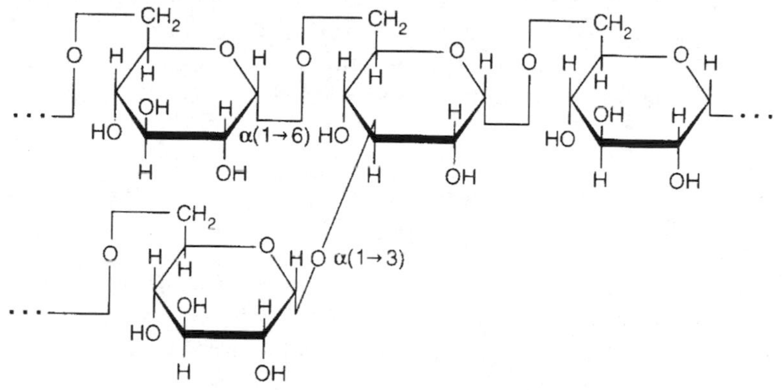

Figure S3-1   Carbohydrate Structures. The structures of the carbohydrates (a) gentiobiose, (b) raffinose, and (c) a portion of a dextran. See Problem 3-10.

3-11.   (a)   A distinguishing feature is the presence of phosphodiester bonds in the DNA but not in the protein. Either (i) digest (hydrolyze) the two molecules chemically and assay for inorganic phosphate (P_i) or (ii) determine whether the molecule can be digested (hydrolyzed) by the enzyme phosphodiesterase.

(b)   A distinguishing feature is the presence of thymidine deoxyribonucleotide in DNA but not in RNA. Hydrolyze the two samples and assay for either (i) the purine thymidine or (ii) the pentose deoxyribose.

(c)   A distinguishing feature is the presence of rigid microfibrils of β-D-glucose in cellulose but not in starch. Either (i) examine the polymer with an electron microscope to look for cellulose microfibrils (see Figure 3-25 on p. 65 of the textbook) or (ii) assay for the enzyme amylase, which can digest $\alpha(1\rightarrow4)$ glycosidic bonds but not $\beta(1\rightarrow4)$ glycosidic bonds and which will therefore digest starch but not cellulose.

(d)   A distinguishing feature is the presence of $\alpha(1\rightarrow6)$ bonds in amylopectin but not in amylose. Assay for either (i) the presence of $\alpha(1\rightarrow6)$ bonds directly or (ii) sensitivity of the polymer to digestion by an enzyme that can hydrolyze $\alpha(1\rightarrow4)$ but not $\alpha(1\rightarrow6)$ bonds (or vice versa).

(e)   A distinguishing feature is the presence of four subunits in hemoglobin but not in myoglobin. Either (i) determine the molecular weight of the native (intact) protein (by ultracentrifugation, most likely) and compare it with the known molecular weights of hemoglobin and myoglobin or (ii) subject the native protein to denaturing conditions to determine whether it will dissociate into multiple subunits or not.

(f)   A distinguishing feature is the presence of glycerol but the absence of phosphorus in the case of the triacylglycerol. Hydrolyze the two samples and assay for the presence of (i) glycerol and (ii) inorganic phosphate.

(g)   A distinguishing feature is the presence on the glycolipid of a carbohydrate side chain instead of a phosphate group. Hydrolyze the two samples and assay for the presence of (i) carbohydrate and (ii) inorganic phosphate.

(h)   A distinguishing feature is that chitin is a polymer consisting of a single repeating group (GlcNAc), whereas a bacterial cell wall polysaccharide consists of two repeating units (GlcNAc and MurNAc). Hydrolyze the two samples and assay for either (i) the presence of two rather than one digestion product (by chromatography, most likely) or (ii) the presence of MurNAc (by chemical assay, most likely).

3-12.   Numerous different answers are possible because of the wide variety of individual proteins in cells. A few of these answers are given below:

(a)   *Hexokinase*, adds phosphate groups to six-carbon sugars, found in cells using glucose as an energy source; *DNA polymerase*, links nucleotides together to produce DNA, found in dividing cells; *protease*, breaks proteins down into amino acids, produced by cells of the digestive system.

(b)   *Collagen*, a fibrous protein giving strength to tissues, found in cells of connective tissue such as tendons; *keratin*, a fibrous protein giving stiffness to hair, found in hair-producing epidermal cells in mammals.

(c)   *Actin*, a component of microfilaments, found in muscle cells and in the cytoskeleton of many cell types; *tubulin*, a component of microtubules, found in dividing eukaryotic cells where it helps pull chromosomes apart; *flagellin*, a component of bacterial flagella, found in motile bacteria.

(d)   *Transcription factors* bind to DNA sequences to turn genes on, and are found in nearly all cells; *lac repressor* binds to DNA to turn off genes encoding enzymes needed for lactose utilization, and is found in bacteria; *Myc protein*, a transcription factor that stimulates cell proliferation, is overly active in some cancer cells.

(e)   *GLUT*, glucose transporter, is found in cells that import glucose (such as red blood cells); *potassium channel*, transports K^+ ions, found in neurons where it helps maintain potassium ion gradient.

(f)   *Insulin* helps cells take up and use glucose and is produced by pancreatic cells when blood glucose increases; *glucagon* causes liver cells to produce glucose, and is produced by the pancreas when blood glucose drops.

(g)   *Insulin receptor* binds insulin to initiate glucose utilization, and is found in cells requiring glucose; *acetylcholine receptor* binds the neurotransmitter acetylcholine to initiate nerve transmission, and is found in neurons.

(h)   *Antibodies*, globular proteins that recognize microorganisms, are found in white blood cells; *chitinase* degrades the fungal cell wall, and is found in certain plant cells.

(i)   *Ferritin*, a protein that binds and stores iron, is found in almost all cells; *gliadin*, a storage protein that is a source of amino acids in seeds, is found in kernels of wheat and other cereals.

3-13.   (a)   The two polymers have very different properties because they differ in structure as a result of differing 1→4 linkages between glucose monomers. The $\beta(1{\rightarrow}4)$ linkage of cellulose gives it a rigid, fibrous structure not seen in starch, which has an $\alpha(1{\rightarrow}4)$ linkage between glucose monomers.

(b)   The rigid, fibrous structure of cellulose and its virtual insolubility in water make this polymer a suitable component of the plant cell wall, and the more flexible structure of starch and its greater solubility in water make this polymer a suitable storage macromolecule.

3-14.   (a)   A lipid is a molecule that is preferentially soluble in an organic solvent rather than in water. This definition is therefore based on the solubility properties of the molecule rather than on the chemical nature of the subunits or the bond that links subunits together, as is the case for proteins, nucleic acids, and carbohydrates.

         (b)   Phosphatidyl choline > Fatty acid > Triacylglycerol > Estradiol > Cholesterol

         (c)   The presence of an oleate will introduce a bend in one of the side chains that will closely approximate the shape of sphingomyelin, because the latter also has a double bond and hence a bend.

         (d)

−11°C: Linolenic acid	+63°C: Palmitic acid
+5°C: Linoleic acid	+70°C: Stearic acid
+16°C: Oleic acid	+76.5°C: Arachidic acid

         (e)   Phosphatidyl serine: the phosphoserine group

Sphingomyelin: the phosphocholine group

Cholesterol: the hydroxyl group (only slightly hydrophilic)

Triacylglycerol: the ester bonds between glycerol and the fatty acids.

3-15.   (a)   Partial hydrogenation results in the partial reduction of C=C double bonds to C–C single bonds in the fatty acyl chains of the triglycerides.

         (b)   Before partial hydrogenation, the shortening was vegetable oil and therefore liquid at room temperature.

         (c)   Partial hydrogenation makes the shortening solid instead of liquid and allows it to be used as a substitute for animal fat.

         (d)   Although the ingredients for making the shortening may be 100% polyunsaturated oils, to convert them into solid shortening they must be hydrogenated and thus become saturated fats in the final product.

CHAPTER

4

Cells and Organelles

4-1.   (a)   Archaea are ancient bacteria but are not the ancestors of modern bacteria. Bacteria first diverged from a primitive cell lineage that later branched into both Archaea and Eukarya.

(b)   Bacteria differ from eukaryotes in having no nucleus, mitochondria, or chloroplasts, but, like all cells, have ribosomes. However, prokaryotic ribosomes differ in size and protein composition from eukaryotic ribosomes.

(c)   It is true that animal cells have an extracellular matrix for support instead of a cell wall, but this is not true of all eukaryotic cells. Plants, algae, and fungi are eukaryotic cells that do have cell walls, which provide support.

(d)   In a typical human muscle cell, the cytoplasmic ribosomes are the eukaryotic type, and the mitochondrial ribosomes are the prokaryotic type.

(e)   DNA is found not only in the nucleus of a eukaryotic cell but also in the mitochondria and, if present, in the chloroplasts.

(f)   Bacteria do not have organelles but carry out ATP synthesis or photosynthesis using the plasma membrane.

(g)   Not all infectious agents have DNA, RNA, and protein. Some viruses have RNA and protein only, viroids have only RNA, and prions are composed only of protein.

(h)   Although some DNA in cells has been called "junk DNA," it may simply be that we do not yet understand its function.

4-2.   (a)   Secretion; pancreatic cells synthesize a variety of digestive enzymes and hormones, which are then secreted into the intestinal tract (enzymes) or the bloodstream (hormones).

(b)   Motility; muscle cells are capable of contraction, using the energy of ATP to cause movement.

(c)   Photosynthesis; palisade cells are the site of much photosynthetic activity in the leaf.

(d)   Absorption; intestinal mucosal cells are especially well suited for this function because they contain microvilli that greatly increase the absorptive surface area of the cell (see Figure 4-2 on p. 77 of the textbook).

(e)   Transmission of electrical impulses; the most important function of nerve cells is to conduct electrical signals from one part of the body to another.

(f)   Cell division; of the cell types listed, only bacterial cells are capable of rapid and repeated division, with only about 20–30 minutes between division under optimal conditions in at least some species.

4-3.   (a)   Ionic bonds between oppositely charged regions of the molecules, hydrogen bonds between donors and acceptors, hydrophobic bonds between nonpolar groups, and van der Waals interactions between closely spaced atoms.

(b)   High temperature will break the weak hydrogen bonds and denature the proteins. High salt will interfere with ionic bonding, and extremes of pH can change the charge on acidic and basic residues of the proteins, interfering with both ionic bonding and hydrogen bonding.

(c)   You could sequence the rRNA to determine the source organism. For the ribosomal proteins, you could sequence the proteins themselves (difficult) or the genes that encode them (easy).

(d)   You would need to add amino acids, tRNAs, amino acyl-tRNA synthetases, and a source of ATP.

4-4.   (a)   look for plastids or a large vacuole.

(b)   become flaccid as water is drawn out of its cells.

(c)   a ribosome, microtubule, microfilament, etc.

(d)   salt water, hot springs, acidic environments, and sulfur containing environments.

(e)   they are similar in size.

(f)   either DNA or RNA but not both.

4-5.   (a)   Look for different types of ribosomal RNAs, a defining difference between bacteria and archaea. Archaea also may have histone-like proteins associated with their DNA, branched polyisoprenoids in the cell membrane, or cell walls containing protein. Bacteria and archaea also differ in their mechanisms of transcription and translation.

(b)   Look for the presence of ribosomes, which are found on rough ER but not on smooth ER.

(c)   Analyze for the presence of any of the enzymes of photorespiration, which are present in leaf peroxisomes but not in animal peroxisomes.

(d)   Look for the presence of catalase (peroxisomes) versus lysozyme and other acid hydrolytic enzymes (lysosomes).

(e)   Look for the presence of a capsid (protein coat); a virus has an external protein layer, whereas a viroid does not.

(f)   Analyze for the presence of actin, which is the main protein of microfilaments but is not a component of intermediate filaments. (Alternatively, you might measure the diameter of the filament, but that is not likely to be a definitive means of distinguishing between microfilaments and intermediate filaments, because they do not differ very much in diameter.)

(g)   Analyze for the chemical nature of the nucleic acid (e.g., for the presence of ribose or deoxyribose); polio virus contains RNA as its genetic information, whereas herpes simplex virus is a DNA-containing virus.

(h)   Measure the sedimentation coefficient, which is about 70S for prokaryotic ribosomes and 80S for eukaryotic ribosomes.

(i)   Compare the sizes of the RNAs. miRNA is regulatory and will be much shorter, usually only about 20 nucleotides in length. Because mRNA will be long enough to encode a protein, it will typically contain hundreds of nucleotides.

4-6.   (a)   B   (d)   N   (g)   A

           (b)   N   (e)   A   (h)   N

           (c)   B   (f)   N   (i)   B

4-7.   (a)   3   (c)   5   (e)   4   (g)   2

           (b)   7   (d)   1   (f)   6

4-8.   (a)   Mitochondrion; underactive due to "poisoning" of the electron transport system by cyanide (which binds to the terminal oxidase in the system, preventing transfer of electrons to oxygen).

           (b)   Peroxisome; underactive due to the absence or inactivity of enzymes involved in the catabolism of fatty acids with very long hydrocarbon chains.

           (c)   Nucleus; overactive in mitotic cell division.

           (d)   Microtubule; underactive because of defective protein components, which prevent or reduce the motion of the flagella that are responsible for the motility of sperm cells; hence, the sperm cells' effectiveness in reaching and fertilizing an egg cell is compromised.

           (e)   Lysosome; underactive due to the absence of the hydrolase enzyme and the consequent lack of ganglioside degradation.

           (f)   Extracellular enzyme; underactive (or inactive) intestinal lactase (the enzyme that hydrolyzes lactose to glucose and galactose) means that lactose cannot be completely digested and absorbed in the small intestine and is converted by bacteria in the large intestine into toxic products that cause abdominal cramps and diarrhea. Eliminating milk from the diet usually alleviates these symptoms. (Lactose intolerance is common among adults of most human races except Northern Europeans and some Africans. It results from the disappearance after childhood of lactase from the intestinal cells.)

4-9.   (a)   Viruses resemble cells in the following ways: they are too small to see with the naked eye; they are composed primarily of carbon, hydrogen, and oxygen; they contain nucleic acids and proteins; they can cause disease; they (sometimes) have a membrane covering.

(b)   Viruses differ from cells in the following ways: they are much smaller than most cells; they do not have both DNA and RNA; they cannot replicate on their own; they do not make their own membranes; they have, at most, one or a few enzymes; they do not have cytoplasm or ribosomes.

(c)   Arguments in favor of viruses being considered alive could use the points presented in the answer to Problem 4-9a. Also, they can be infectious, can multiply, can inherit genetic material, and can direct protein synthesis using their nucleic acids. Arguments in favor of viruses being considered not alive could use the points presented in the answer to Problem 4-9b. In addition, they have no metabolism outside the host cell, do not move, and do not respond to their environment.

(d)   One reason that viral illnesses are more difficult to treat is because viruses have fewer components and thus fewer targets than bacteria. Typical antibiotics target the bacterial cell wall, ribosomes, or enzymes—targets that are not found in viruses. Also, stopping viral multiplication may require inhibiting the metabolism of the host cell, which often has side effects for the host organism.

(e)   Find a virus-specific molecule (capsid protein) or enzyme (reverse transcriptase in HIV) not found in the host and target that molecule. Alternatively, find a process required for viral multiplication (protease processing of the capsid proteins) that is not needed for host cell survival and block that process.

5

Bioenergetics:
The Flow of Energy
in the Cell

5-1.   (a)   (1.94 cal/min · cm^2) (5.26 × 10^5 min/year) (1.28 × 10^{18} cm^2)
= **1.3 × 10^{24} cal/year.**

      (b)   Some incident radiation is reflected back into space and much is absorbed by components of the Earth's atmosphere. Atmospheric ozone plays an important role in filtering out ultraviolet radiation, whereas water vapor is responsible for most of the absorption in the infrared range.

      (c)   Much of the radiation falls on areas of the Earth's surface where the climate is not favorable (too hot, too cold, too dry) for growth of phototrophic organisms during at least part of the year. In addition, about two-thirds of the Earth's surface is covered by oceans that, though quantitatively significant in global photosynthesis, have in general only a very low density of phototrophic organisms and therefore a low efficiency of light utilization. Moreover, incident radiation represents a broad spectrum of wavelengths that can be used with varying degrees of efficiency by photosynthetic pigments, as discussed further in Chapter 11 of the textbook.

5-2.   (a)   Glucose is 40% carbon (72/180 = 0.4), so 5 × 10^{16} g carbon represents about **12.5 × 10^{16} g organic matter.** (That is about 140 billion tons, if nonmetric units help you imagine the magnitude of the process.)

      (b)   (12.5 × 10^{16} g)(3.8 kcal/g) = 4.75 × 10^{17} kcal = **4.75 × 10^{20} cal.**

      (c)   (4.75 × 10^{20} cal)/(1.3 × 10^{24} cal) = 3.7 × 10^{-4} = **0.037%.** (This means that more total solar energy will be received by the earth during the next 12 months than has been trapped photosynthetically since 700 BC!)

      (d)   Virtually all of it must be consumed by chemotrophs, because the earth is not undergoing any dramatic annual increase in amount of phototrophic organic matter accumulated.

5-3.   (a)   Use of blood sugar as a source of energy for muscle contraction; important for animal motility.

      (b)   Use of chemical energy to cause flash of light by firefly; important as mating signal.

(c)   Photosynthetic use of sunlight to synthesize sugar; entire biosphere depends on this process as its energy link with the sun.

(d)   Use of chemical energy to generate a potential and deliver an electric shock; defense mechanism of electric eel.

(e)   Use of chemical energy to pump protons into stomach and maintain low pH; aid to digestion of foodstuffs.

5-4. (a)   The ΔH value of a reaction is a measure of the difference in heat content between the reactants and the products of the reaction and, therefore, of the heat exchange that accompanies the reaction under specified conditions. A negative ΔH value means that heat is liberated as the reaction proceeds. The ΔG value of a reaction is a measure of the difference in free energy between the reactants and products of the reaction and, therefore, of the exchange of free energy that accompanies the reaction under specified conditions. A negative ΔG value means that free energy is liberated as the reaction proceeds (and that the reaction is therefore thermodynamically spontaneous in the direction written).

(b)   The change in free energy, ΔG, is the algebraic sum of the change in enthalpy and the change in entropy that occur as reactants are converted into products under specified conditions. In other words, ΔG takes into account both the difference in heat content and the difference in entropy between the reactants and products, whereas ΔH measures only the former.

(c)   The reaction involves the conversion of a complex, 24-atom organic molecule into 12 simpler 3-atom molecules (with the additional 12 atoms provided by water). The distribution of carbon, hydrogen, and oxygen atoms is therefore much more random after the reaction than before. As a result, the entropy of the products is greater than that of the reactants, so the entropy change is *positive*.

(d) $\Delta G = \Delta H - T \, \Delta S$

$\Delta S = (\Delta H - \Delta G)/T$

$= (-673 - [-686])/(25 + 273) = +13/298 = +0.0436$

$= +43.6$ **cal/mol-K.**

The sign of ΔS agrees with the prediction in part c.

(e)   Because the overall reaction of photosynthesis is the reverse of Reaction 5-24 in the textbook, the values of ΔS, ΔH, and ΔG will be identical in magnitude to the corresponding values for Reaction 5-24, but opposite in sign.

5-5. (a) $K'_{eq} = [F6P]/[G6P] = 0.5$

The concentrations of F6P and G6P after an overnight incubation in the presence of the enzyme catalyst will be:

[F6P] = xM [G6P] = $(0.15 - x)M$

We can therefore write

$K'_{eq} = (x)/(0.15 - x) = 0.5$

Thus, $1.5x = 0.075$ and $x = 0.075/1.5 = 0.05 \, M$

Given that the F6P concentration is 0.05 mol/L or 0.05 mmol/mL, 10 mL of solution will contain **0.5 mmol of F6P.**

(b)   The same answer; the equilibrium concentrations of products and reactants depend only on the value of K'_{eq} and the temperature, not on the starting concentrations.

(c)   In the absence of a catalyst, the reaction would not proceed to any measurable extent, so the F6P concentration after the incubation would be negligible.

(d)   No, not without knowing the effect of temperature on the K'_{eq} value.

5-6.   (a) $\Delta G^{\circ\prime} = -RT \ln K'_{eq} = (-1.987)(298) \ln (0.165) = -592(-1.802)$

$= +1067 \text{ cal/mol.}$

To convert 1 mole of 3-phosphoglycerate to 2-phosphoglycerate under conditions where the concentrations of both are maintained constant at 1.0 M would require the input of 1067 cal of free energy; the reaction is endergonic under standard conditions.

(b) $\Delta G' = \Delta G^{\circ\prime} + (1.987)(298) \ln \dfrac{4.3 \times 10^{-6}}{61 \times 10^{-6}}$

$= +1067 + 592 \ln (0.0705) = 1067 + 592(-2.652)$

$= -500 \text{ cal/mol.}$

To convert 1 mole of 3-phosphoglycerate to 2-phosphoglycerate under the prevailing concentrations in the red blood cell will result in the liberation of 500 cal of free energy; the reaction is exergonic under prevailing conditions. $\Delta G'$ and $\Delta G^{\circ\prime}$ differ because $\Delta G^{\circ\prime}$ measures the energetics of the reaction under standard conditions, whereas $\Delta G'$ measures the energetics of the reaction under actual, or prevailing conditions.

(c)   The concentration of 2-phosphoglycerate can rise only to the equilibrium value. If it goes above that, equilibrium will lie to the left, and the reaction will proceed in the reverse direction.

To calculate the equilibrium value:

$K'_{eq} = 0.165 = \dfrac{[2-\text{phosphoglycerate}]}{61 \times 10^{-6}}$

$[2\text{-phosphoglycerate}] = (0.165)(61 \times 10^{-6}) = 10 \times 10^{-6} \, M$

$= 10 \, \mu M.$

(Alternatively, you can set $\Delta G'$ equal to zero and solve for [2-phosphoglycerate] using Equation 5-21 on p. 122 of the textbook.)

5-7. (a) The $\Delta G^{\circ\prime}$ value is positive, so the K'_{eq} value is less than one:

$\Delta G^{\circ\prime} = (1.987)(298) \ln K'_{eq} = -1800$ cal/mol.

Thus, $\ln K'_{eq} = -1800/592 = -0.304$, so $K'_{eq} = 0.048$.

The equilibrium therefore lies to the **left**.

(b) Because the $\Delta G^{\circ\prime}$ value is the value of $\Delta G'$ under standard conditions, the reaction will tend to proceed to the **left**. The $\Delta G'$ value in that direction is **–1.8 kcal/mol**.

(c) $\Delta G' = \Delta G^{\circ\prime} + RT \ln ([G3P]/[DHAP])$

$= +1800 + (1.987)(298) \ln (0.01) = +1800 + 592(-4.605) = 1800 - 2726$

$= -926$ cal/mol $= $ **–0.93 kcal/mol**.

(d) Written in the direction required for the Calvin cycle, the reaction becomes

$\qquad$ G3P $\rightleftharpoons$ DHAP $\qquad\qquad \Delta G^{\circ\prime} = -1800$ cal/mol

To calculate the ratio $X = [DHAP]/[G3P]$ at which the $\Delta G'$ will be -3000 cal/mol in this direction, we can write

$-3000 = -1800 + RT \ln X$

So $592 \ln X = -3000 + 1800 = -1200$ cal/mol

$\ln X = 1200/592 = -2.027$ and $X = $ **0.132.**

Thus, the [DHAP]:[G3P] ratio must be at or less than 0.132. If it is higher than this limit, the value of $\Delta G'$ will be less negative than -3000 cal/mol. Conversely, the [G3P]:[DHAP] ratio must be at least $1/0.132 = $ **7.58.**

5-8. (a) Because $\Delta G^{\circ\prime} = 0$, the equilibrium constant must be 1.0. The reaction will therefore continue to the right until all species are present at equimolar concentrations of 0.005 M each.

(b) Assume x mol/liter of succinate react with x mol/liter of FAD to generate x mol/liter each of fumarate and FADH$_2$.

At equilibrium:

[succinate] $= 0.01 - x$

[FAD] $= 0.01 - x$

[FADH$_2$] $= 0.01 + x$

[fumarate] $= x$

$$K'_{eq} = 1.0 = \frac{[\text{fumarate}][\text{FADH}_2]}{[\text{succinate}][\text{FAD}]} = \frac{(x)(0.01+x)}{(0.01-x)(0.01-x)}$$

$$= \frac{0.01x + x^2}{0.0001 - 0.02x + x^2}$$

$0.01x + x^2 = 0.0001 - 0.02x + x^2$, so $0.03x = 0.0001$,

and therefore $x = 0.0033$.

Equilibrium concentrations are therefore:

[succinate] = 0.0067 M

[FAD] = 0.0067 M
[FADH$_2$] = 0.0133 M

[fumarate] = 0.0033 M.

(c) $\Delta G' = \Delta G^{\circ\prime} + 592 \ln \dfrac{[\text{fumarate}][\text{FADH}_2]}{[\text{succinate}][\text{FAD}]} = -1500 \text{ cal/mol}$

$$= 0 + 592 \ln \frac{(2.5 \times 10^{-6})(5)}{[\text{succinate}]} = -1500 \text{ cal/mol}$$

$$\ln \frac{(12.5 \times 10^{-6})}{[\text{succinate}]} = \frac{1500}{592}$$

so $\dfrac{(2.5 \times 10^{-6})}{[\text{succinate}]} = e^{-2.534} = 0.07934$

$12.5 \times 10^{-6} = 0.07934 \,[\text{succinate}]$

so [succinate] = $(12.5 \times 10^{-6})/0.07934 = 157 \times 10^{-6}$ M = **157 μM.**

5-9. (a) Protein folding occurs spontaneously, which means that it is a thermodynamically spontaneous process, so the value for ΔG must be *negative*. This in turn means that the ΔG value for protein unfolding, or denaturation, must be *positive* because the value of a thermodynamic parameter for the reverse reaction always has the same numeric value but the opposite sign.

(b) Protein folding results in greater order, or less disorder, so the ΔS value will be *negative*. That means, in turn, that the ΔS value for denaturation will be *positive*.

(c) The contribution of ΔS to the free energy change will be positive for protein folding because the equation for ΔG includes the term $(-T\Delta S)$. T is obviously positive, so if ΔS is negative, the term $(-T\Delta S)$ will be positive.

(d) The main covalent bond that must be broken in the denaturation of many proteins is the disulfide bond. Noncovalent bonds and interactions that must be broken include hydrogen bonds, ionic bonds (also called electrostatic interactions), van der Waals interactions, and hydrophobic interactions (see Figure 3-5 on p. 45 of the textbook). Heating adds energy to the system, making it more difficult for the noncovalent bonds to maintain the polypeptide in its native conformation. Extremes of pH cause changes in the ionization or protonation state of ionizable functional groups, thereby disrupting both ionic and hydrogen bonds.

5-10. (a) The K_{eq} for the transport of an uncharged solute across a membrane is always 1 because if the solute is allowed to come to equilibrium across the membrane, the concentrations on both sides of the membrane will be identical. And with $K_{eq} = 1$, $\Delta G°$ will be zero because it is defined as $-RT \ln K_{eq}$ and the ln of 1.0 is zero.

(b) The respective equations for inward and outward transport of an uncharged solute S are as follows:

For inward transport: $\Delta G_{inward} = +RT \ln [S]_{inside}/[S]_{outside}$

For outward transport: $\Delta G_{outward} = +RT \ln [S]_{outside}/[S]_{inside}$

(c) The concentration of lactose inside the cell is much greater (by a factor of 50, in fact) compared to the lactose concentration on the outside, so we can confidently predict that the inward movement of lactose will be nonspontaneous and that the ΔG_{inward} for lactose will therefore be positive.

(d) $\Delta G_{inward} = +RT \ln [S]_{inside}/[S]_{outside}$

$= +(1.987)(273 + 25) \ln (0.010\ M)/(0.0002\ M)$

$= +592 \ln (50) = +592(3.912) - +2316$ cal/mole $= $ **+2.32 kcal/mole**

The calculated value of ΔG_{inward} confirms the prediction in part (c).

(e) At –7.3 kcal/mole, the energy yield of ATP hydrolysis is more than adequate to drive the uptake of one lactose molecule transported per molecule of ATP hydrolyzed. In fact, ATP hydrolysis could drive the uptake of two or even three lactose molecules per ATP and still be spontaneous: $-7.3 + 3(2.32) = $ **–0.38 kcal/mole.** Moreover, the value of ΔG under cellular conditions is typically much more negative than $-\Delta G°$; it is often in the range –10 to –12 kcal/mole, in fact.

5-11. (a) For the overall reaction A → D,

$$K'_{AD} = \frac{[D]}{[A]} = \frac{[D][B][C]}{[A][B][C]} = \frac{[B][C][D]}{[A][B][C]} = K'_{AB} \cdot K'_{BC} \cdot K'_{CD}.$$

(b) $\Delta G°'_{AD} = RT \ln K'_{AD} = -RT \ln [K'_{AB} \cdot K'_{BC} \cdot K'_{CD}]$

$= -RT \ln K'_{AB} -RT \ln K'_{BC} -RT \ln K'_{CD} = \Delta G°'_{AB} + \Delta G°'_{BC} + \Delta G°'_{CD}.$

(c) $\Delta G^{\circ\prime}_{AD} = \Delta G^{\circ}_{AD} + RT \ln \dfrac{[D]}{[A]} = \Delta G^{\circ}_{AD} + RT \dfrac{[D][B][C]}{[A][B][C]}$

$= \Delta G^{\circ}_{AD} + RT \ln \dfrac{[B][C][D]}{[A][B][C]}$

$= \Delta G^{\circ}_{AB} + \Delta G^{\circ\prime}_{BC} + \Delta G^{\circ\prime}_{CD} + RT \ln \dfrac{[B]}{[A]} + RT \ln \dfrac{[C]}{[B]} + RT \ln \dfrac{[D]}{[C]}$

$= \Delta G^{\circ}_{AB} + RT \ln \dfrac{[B]}{[A]} + \Delta G^{\circ\prime}_{BC} + RT \ln \dfrac{[C]}{[B]} + \Delta G^{\circ}_{CD} + RT \ln \dfrac{[D]}{[C]}$

$= \Delta G^{\circ}_{AB} + \Delta G^{\circ\prime}_{BC} + \Delta G^{\circ\prime}_{CD}.$

5-12.   (a)

glucose + $P_i \rightarrow$ G6P + H_2O	$\Delta G^{\circ\prime}$ = +3.3 kcal/mol
+ $\underline{ATP + H_2O \rightarrow ADP + P_i}$	$\Delta G^{\circ\prime}$ = –7.3 kcal/mol
glucose + ATP $\rightarrow$ G6P + ADP	$\Delta G^{\circ\prime}$ = +3.3 – 7.3 = **–4.0 kcal/mol**

(b)

P-creatine + $H_2O \rightarrow$ creatine + P_i	$\Delta G^{\circ\prime}$ = –10.3 kcal/mol
+ $\underline{ADP + P_i \rightarrow ATP + H_2O}$	$\Delta G^{\circ\prime}$ = +7.3 kcal/mol
P-creatine + ADP $\rightarrow$ creatine + ATP	$\Delta G^{\circ\prime}$ = –10.3 + 7.3 = **–3.0 kcal/mol**

6

Enzymes:
The Catalysts of Life

6-1. (a) It means that the molecules that ought to react would release energy if they were to do so, but they do not possess enough energy to collide in a way that allows the reaction to be initiated.

(b) Touching a match to a sheet of paper is an example. Thermal activation imparts sufficient kinetic energy to the molecules such that the proportion of them that possess adequate energy to collide and react increases significantly. Once initiated, the reaction is self-sustaining, because reacting molecules release sufficient energy to energize and activate neighboring molecules for reaction.

(c) Interaction of molecules is facilitated (by positioning on the catalyst surface, for example), thereby requiring less energy to activate each molecule and so ensuring that substantially greater numbers of molecules possess adequate energy to initiate reaction without any elevation in temperature.

(d) Advantages: specificity, more exacting control.
Disadvantages: much more susceptible to inactivation by heat, extremes of pH, and so on; also, much energy needs to be expended to synthesize the enzyme molecules.

(e) This can happen by means of a mechanism known as quantum tunneling in which a hydrogen atom used as a substrate by dehydrogenases can move through the activation barrier rather than over it.

6-2. (a) The activation energy diagram for the catalase reaction is shown in Figure S6-1. To say that platinum lowers the activity energy from 18 to 13 kcal/mol means that platinum is a surface on which molecules of hydrogen peroxide are positioned such that their reaction to form water and molecular oxygen is more favorable than when hydrogen peroxide is free in solution. Specifically, it means that a pair of H_2O_2 molecules on a platinum surface requires only about 72% as much activation energy to initiate a reaction between them as when they are present in solution. As a result, many more of the H_2O_2 molecules possess adequate energy to interact at the prevailing temperature. Catalase provides a surface that is even more conducive to reaction; a pair of H_2O_2 molecules at the active site of catalase requires only about 39% as much activation energy as when the molecules are present in solution.

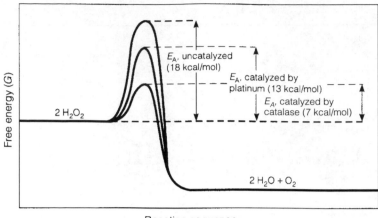

Figure S6-1 Effect of Catalysis on the Activation Energy. The activation energy E_A for the decomposition of H_2O_2 to H_2O and O_2 is about 18 kcal/mol. This value is reduced to about 13 kcal/mol by platinum, a metallic catalyst, and to about 7 kcal/mol by the enzyme catalase. See Problem 6-2a.

(b)   The active site of the enzyme provides a surface that is very specific for the binding of H_2O_2 molecules. The particular amino acids and the iron–porphyrin complex that are present at the active site create a chemical environment with the right functional groups and pH to favor maximally the binding of H_2O_2 molecules and their subsequent reaction to form the desired products.

(c)   The rate of hydrogen peroxide decomposition can also be accelerated by an increase in temperature. However, this is not an appropriate means of increasing reaction rates within cells because most cells can function only within a rather limited temperature range; for most cells, increases in temperature lead to the denaturation of proteins and hence to cellular dysfunction or death.

6-3.   (a)   To decompose the same quantity of H_2O_2 in the presence of an equivalent amount of ferric ions would take $10^8/(3 \times 10^4)$ or 3.33×10^3 times longer:

$3.33 \times 10^3 \times 1$ minute $\times$ 1 hour/60 min = **55.5 hours.**

(b)   In the absence of a catalyst, the process would take 10^8 times longer:

$10^8 \times 1$ minute $\times$ 1 hour/60 min $\times$ 1 day/24 hour $\times$ 1 year/365.25 days = **190 years.**

(c)   It should be clear that having to wait hours for a chemical reaction to occur to a significant extent is not compatible with the rapid changes that cells must be able to effect. Hundreds of years is in the realm of the ridiculous!

6-4. (a) Figure 6-4a on p. 134 of the textbook illustrates the temperature dependence of a typical human enzyme and a typical enzyme from a thermophilic bacterium. Each has a temperature optimum that is at or near the temperature of the organism (37°C for the human body, about 75°C for a typical hot spring). Clearly, the bacterial enzyme must be much more heat stable, presumably due to the number and strength of both noncovalent bonds (hydrogen bonds and electrostatic interactions) and covalent (disulfide) bonds. In both cases, the reaction velocity increases steadily as the temperature is increased from a colder temperature to the temperature optimum of the enzyme, consistent with the effect of temperature on chemical reactions in general, which usually double in reaction velocity for every 10°C increase in temperature. As the temperature is raised above the optimum, however, the enzyme undergoes thermal denaturation resulting in loss of activity.

Figure 6-4b in the textbook illustrates the pH dependence of two enzymes with very different pH optima. The pH optimum for an enzyme corresponds to the proton concentration at which ionizable groups on the enzyme and the substrate molecules are in the most favorable form for chemical reactivity. Changes in pH away from the optimum result in loss of enzyme activity due to ditration of the ionizable groups on the enzyme or the substrate.

(b) For Figure 6-4a, both enzymes are maximally active at or near the temperature of the milieu in which they are found—the human body in one case, a thermal hot spring in the other. For Figure 6-4b, the difference in pH optima for the two enzymes reflects the very different environments in which the two enzymes are active (the stomach for pepsin, the small intestine for trypsin).

(c) An enzyme with a very flat pH profile probably has no amino acids at its active site that undergo ionization or protonation, and it probably catalyzes a reaction in which neither the substrates nor the products can be ionized or protonated.

6-5. (a) C   (c) B   (e) C
      (b) A   (d) B   (f) A

6-6. (a) If product accumulation becomes significant, the back-reaction may begin to occur, and the net activity in the forward direction will be correspondingly reduced.

(b) See the graph in Figure S6-2 (solid line). Doubling of substrate concentration always results in less than a twofold increase in velocity because the curve follows a hyperbolic equation, such that the relationship between substrate concentration and velocity is not linear at any point.

(c) The data for the double-reciprocal plot are calculated as shown in Table S6-1. For the double-reciprocal plot, see the graph in Figure S6-3 (solid line).

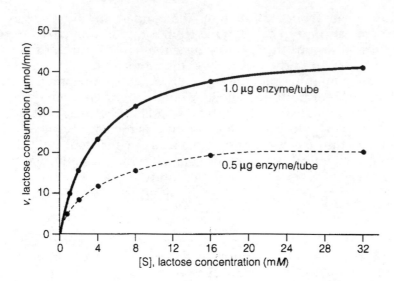

Figure S6-2   Kinetics of the β-Galactosidase Reaction. The dependence of initial reaction velocity on substrate concentration is shown for the data of Problem 6-6 with two concentrations of enzyme: 1.0 μg per tube (solid line), and 0.5 μg per tube (dashed line).

Table S6-1. Double-Reciprocal Data for Problem 6-6.

$\dfrac{[S]}{mM}$	$\dfrac{1/[S]}{1/mM}$	$\dfrac{v}{\mu mol/min}$	$\dfrac{1/v}{min/\mu mol}$
1.0	1.000	10.0	0.100
2.0	0.500	16.7	0.060
4.0	0.250	25.0	0.040
8.0	0.125	33.3	0.030
16.0	0.0625	40.0	0.025
32.0	0.03125	44.4	0.022

(d) K_m: X-intercept is –0.25, so $K_m = -1/-0.25 = $ **+4.0 mM.**

V_{max}: Y-intercept is 0.02, so $V_{max} = 1/0.02 = $ **50 μmol/min.**

(e)   Results with 0.5 μg enzyme per tube are shown as the dashed line on the graph of Figure S6-2. Half as much enzyme results in half the total rate of lactose hydrolysis at any concentration of lactose (that is, reaction velocity is linear with enzyme concentration).

6-7.   (a)   In a double-reciprocal plot such as Figure 6-20 on p. 154 of the textbook, K_m is the negative reciprocal of the x-intercept: $K_m = -1/(20) = $ **+0.05 mM.**

This value tells us that the enzyme galactokinase will be functioning at 50% of its maximum velocity when galactose is present at a concentration of 0.05 mM.

(b) V_{max} corresponds to the reciprocal of the y-intercept:
$V_{max} = 1/0.1 = $ **10 μmol/min.**

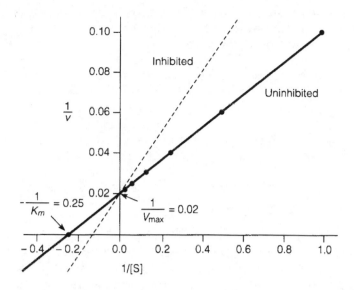

Figure S6-3   Double-Reciprocal Plot for the β-Galactosidase Reaction. Reciprocal values for the initial reaction velocity v and the substrate concentration [S] were determined for the data of Problem 6-6 and plotted as $1/v$ versus $1/[S]$ (solid line). In the presence of a competitive inhibitor that increases the apparent K_m value by a factor of two, the line will be twice as steep (dashed line).

This value tells us that, for the specific amount of the enzyme used in the assay, as the substrate concentration tends toward infinity, the reaction velocity will tend toward (i.e., will asymptotically approach) 10 mmol/min.

(c)   You would expect to get the same V_{max} value because that is the velocity obtained when both substrates are present at high, nonlimiting concentrations, which will be true at the highest substrate concentrations used in both cases.

(d)   The K_m values for galactose and ATP are measures of the affinity of the enzyme for these two substrates. There is no *a priori* reason to assume that the enzyme would have the same affinity for two substrates, especially when they are chemically very different from each other.

6-8.   (a)   Number of moles of enzyme present:

2 μg × 10^{-6} g/μg × 1 mole/30,000 g = $(2 \times 10^{-6})/(3 \times 10^4)$ = 6.67×10^{-11} moles.

Rate of CO_2 consumption:

$(6.67 \times 10^{-11}$ moles$)$ (1×10^6)/sec = 6.67×10^{-5} mole/sec = **0.0667 mmoles/second.**

(b)   One mole of a gas occupies 22.4 liters at STP, so 0.0667 mmoles of CO_2 occupies $(0.0667$ mmoles$)$ × 22.4 mL/mmole = 1.5 mL.

The rate of CO_2 consumption is therefore **1.5 mL/second.**

6-9.   (a)   Because diisopropyl fluorophosphate binds covalently to the hydroxyl group of a specific amino acid residue of the target enzyme it cannot be an allosteric effector, because allosteric effectors bind reversibly.

       (b)   The enzyme hexokinase is inhibited by its own product, glucose-6-phosphate, and is therefore not an example of feedback inhibition, which involves regulation by molecules other than reactants and products.

       (c)   Glycogen synthase, in contrast to glycogen phosphorylase, is inactive in the phosphorylated form and active in the dephosphorylated form.

       (d)   An enzyme that is subject to allosteric activation is most likely to catalyze the first reaction in a biosynthetic pathway.

       (e)   If researchers claim that an enzyme is allosterically activated by compound A and allosterically inhibited by compound B, both of these claims can be correct provided that the enzyme has separate binding sites for the two effectors. (In fact, some regulatory enzymes have binding sites for several activators and several inhibitors. For examples of such enzymes, see Figure 9-12 on p. 244 of the textbook.)

6-10. (a)   Because irreversible inhibitors bind to the enzyme covalently, you could analyze the enzyme molecule after adding the inhibitor to see if there was an increase in molecular weight or charge corresponding to the weight or charge of the inhibitor. You could test whether the inhibitor contains any heavy metals, which often are irreversible inhibitors. You could test whether it is possible to separate the inhibitor from the inhibited enzyme (by dilution, for example)—if not, the inhibitor is likely to be covalently bound and thus irreversible.

       (b)   If adding a large amount of substrate overcomes the inhibition and restores maximum enzyme activity, it is competitive inhibition because the substrate is outcompeting the inhibitor for the active site (V_{max} does not change but K_m increases). If this does not restore enzyme activity, the inhibitor is binding at a separate site and is showing noncompetitive inhibition (K_m is not affected but V_{max} decreases).

6-11. (a)   The "burst of energy" needed for muscle contraction is provided by an increased rate of ATP synthesis by the muscle cells. Because the most common substrate for ATP synthesis in a muscle cell is glucose and glucose is stored in the cell as glycogen, the enzyme glycogen phosphorylase must be activated to break down the glycogen to glucose (see Figure 6-17 on p. 149 of the text). Thus, epinephrine and glucagon, when secreted into the bloodstream (from the adrenal gland and the pancreas, respectively), reach the mitochondria and trigger the activation of glycogen phosphorylase (by converting it from the *b* to the *a* form). The activated enzyme then degrades stored glycogen, releasing glucose molecules that are catabolized to generate the needed ATP.

       (b)   If any molecules of glycogen phosphorylase are in the *a* form in the presence of high concentrations of either ATP or glucose, it makes sense for the enzyme molecules to be inhibited allosterically by the ATP or the glucose, because further glycogen catabolism isn't necessary if either ATP is already present at an adequate concentration for cellular needs or an adequate supply of glucose is already available to generate the needed ATP.

(c)   Formation of the pancreatic enzyme carboxypeptidase by trypsin-activated cleavage of its inactive precursor makes good sense because trypsin is a gastric enzyme, so its appearance in the intestine is almost certainly the result of the movement of partially digested food from the stomach to the small intestine, which is the site of carboxypeptidase activity.

6-12.   The point of departure for the derivation is the simple enzyme-catalyzed reaction of Equation 1 and the Michaelis-Menten model for its mechanism shown in Equation 2:

$$S \xrightarrow{\text{enzyme}} P \tag{1}$$

$$E_f + S \underset{k_2}{\overset{k_1}{\rightleftharpoons}} ES \underset{k_4}{\overset{k_3}{\rightleftharpoons}} + P + E_f \tag{2}$$

The velocity v can be expressed as the rate of disappearance of substrate or the rate of appearance of product:

$$v = -\frac{d[S]}{dt} = +\frac{d[P]}{dt} \tag{3}$$

To derive the dependence of v on [S], we begin by writing equations expressing the rates of change in concentrations of S, ES, and P:

$$\frac{d[S]}{dt} = -k_1[E_f][S] + k_2[ES] \tag{4}$$

$$\frac{d[P]}{dt} = +k_3[ES] - k_4[P][E_f] \tag{5}$$

$$\frac{d[ES]}{dt} = k_1[E_f][S] - k_2[ES] - k_3[ES] + k_4[P][E_f] \tag{6}$$

Because we are confining ourselves to the initial stages of the reaction when [P] is essentially zero, equations 5 and 6 simplify to

$$\frac{d[P]}{dt} = k_3[ES] \tag{7}$$

$$\frac{d[ES]}{dt} = k_1[E_1][S] - k_2[ES] - k_3[ES] \tag{8}$$

$$= k_1[E_f][S] - [k_2 + k_2][ES]$$

To proceed we must now assume the steady state at which the enzyme–substrate complex is being broken down at the same rate at which it is being formed, such that the net rate of change in [ES] *is* zero:

$$\frac{d[ES]}{dt} = k_1[E_f][S] - [k_2 + k_3][ES] = 0 \tag{9}$$

This can be rewritten as

$$k_1[E_f][S] = [k_2 + k_3][ES] \tag{10}$$

Clearly, the total amount of enzyme present, E_t, is simply the sum of the free form E_f plus the amount of complexed enzyme ES:

$$E_t = E_f + ES \tag{11}$$

Hence, $$E_f = E_t - ES \tag{12}$$

which, when substituted into Equation 10, yields

$$k_1[E_t - ES][S] = [k_2 + k_3][ES] \tag{13}$$

or $$k_1[E_t][S] - k_1[ES][S] = [k_2 + k_3][ES] \tag{14}$$

Rearranging this yields

$$k_1[E_t][S] = k_1[ES][S] = [k_2 + k_3][ES] \tag{15}$$

$$= (k_1[S] + k_2 + k_3)\,[ES] \tag{16}$$

which can be rewritten as

$$[ES] = \frac{k_1[E_t][S]}{k_1[S] + k_2 + k_3} \tag{17}$$

This is useful because Equations 3 and 7 tell us that the velocity v is simply $k_3\,[ES]$:

$$v = k_3[ES] = \frac{k_1 k_3[E_t][S]}{k_1[S] + k_2 + k_3} \tag{18}$$

Divide both numerator and denominator by k_1:

$$v = \frac{k_3[E_t][S]}{[S] + ((k_2 + k_3)/k_1)} \tag{19}$$

For a given concentration of enzyme, E_t is a constant, so we can define two kinetic constants as follows:

$$V_{\max} = k_3[E_t] \qquad\qquad K_m = (k_2 + k_3)/k_1$$

Rewritten in these terms, Equation 19 becomes

$$v = \frac{V_{\max}[S]}{[S] + K_m}$$

6-13. (a)

| **Eadie-Hofstee Plot** | **Hanes-Wolff Plot** |

Cross-multiplication of the Michaelis-Menten expression yields

Cross-multiplication of the Michaelis-Menten expression yields

$vK_m + v[S] = V_{max}[S].$

$V_{max}[S] = vK_m + v[S].$

Dividing by $[S]$ gives

Dividing by $v \cdot V_{max}$ gives

$vK_m/[S] + v = V_{max}$, so

$[S]/v = K_m/V_{max} + (1/V_{max})[S].$

$v = V_{max} - K_m (v/[S]).$

(b) Plot v versus $v/[S]$.

Plot $[S]/v$ versus $[S]$.

Y-intercept = V_{max}

X-intercept = $-K_m$

Slope = $-K_m$

Slope = $1/V_{max}$

X-intercept = V_{max}/K_m

Y-intercept = K_m/V_{max}

(c) A typical Hanes-Wolff plot is shown in Figure S6-4. The Hanes-Wolff plot is statistically the most satisfactory means of linearizing kinetic data because it has the data points (i.e., the [S] values) spaced out linearly along the X-axis, as in the original Michaelis-Menten plot. Both of the other plots use the reciprocal of [S] for the X-axis, thereby giving too much weight to values determined at low substrate concentrations—which are the data points that are most prone to error because of the small numbers involved.

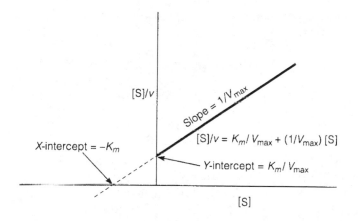

Figure S6-4 The Hanes-Wolff Plot. The ratio $[S]/v$ is plotted as a function of [S]. K_m can be determined from the X-intercept and V_{max} from the slope. See Problem 6-12.

CHAPTER

7

Membranes: Their Structure, Function, and Chemistry

7-1. (a) Localization of function   (f) Regulation of transport

      (b) Intercellular communication   (g) Regulation of transport

      (c) Regulation of transport   (h) Detection of signals

      (d) Localization of function   (i) Intercellular communication

      (e) Localization of function

7-2. (a) Evidence of membrane asymmetry; 1950s.

      (b) Evidence that the membrane permeability of solutes is related to how nonpolar they are; 1890s.

      (c) Evidence that the phospholipid bilayer is not completely covered by surface layers of protein; 1960s.

      (d) Evidence that such particles, when seen on biological membranes, are integral proteins; 1970s.

      (e) Evidence that real membranes are not just phospholipid bilayers; 1920s.

      (f) Evidence of two categories of membrane proteins differing significantly in location within the membrane and in their affinity for an aqueous environment; 1970s.

      (g) Evidence that light-dependent proton pumping can be carried out by a relatively simple protein–pigment complex; in addition, the crystalline array allowed electron diffraction analysis, which led to the first understanding of how a membrane protein is organized within the lipid bilayer; 1970s.

7-3. (a) Because membranes have a hydrophobic interior, polar and charged molecules can pass through membranes only with the help of a protein transporter embedded in the membrane.

      (b) Proteins typically transmit signals from the outside of the cell to the cytoplasm by a change in conformation induced by a molecule binding to the extracellular portion of the protein.

(c)   O-linked and N-linked glycoproteins are formed when sugar chains are attached to the oxygen and nitrogen atoms of the side chains in amino acid residues.

(d)   The three-dimensional structure of a protein can be determined in part even if the protein cannot be isolated from cells in pure form by comparing its nucleotide sequence to that of other genes encoding proteins of known structure. Other biochemical tests involving antibodies or affinity labeling can help determine which regions of the protein are on the surface and which are in the interior.

(e)   You would expect membrane lipids from tropical plants such as palm and coconut to have long-chain saturated fatty acids without C=C double bonds, giving them decreased fluidity so they are functional even in warm temperatures.

7-4.   (a)   If the lipids from 4.74×10^9 erythrocytes have an area of 0.89 m^2 (= 0.89×10^{12} μm^2), then each cell has a monolayer area of $0.89 \times 10^{12} / 4.74 \times 10^9$ = **188 μm^2**. This is almost twice the surface area of an erythrocyte as Gorter and Grendel estimated it at the time, leading to the conclusion that the surface of each cell was covered with two layers (i.e., a bilayer) of lipid.

     (b)   As it turned out, the lipid extraction technique that Gorter and Grendel used was not quantitative, so they underestimated the amount of lipid per cell. In fact, they extracted only about two-thirds of the total lipid from the erythrocytes, so both of their values were off by about the same extent—a classic case of the right conclusion but from flawed data.

7-5.   (a)   If the cell was active in a nonpolar solvent like benzene, its membrane would likely be the reverse of ours. It would have the nonpolar groups on the two surfaces facing the nonpolar solvent, and it would have a hydrophilic interior. The phospholipid head groups, being confined to the center of the membrane, would likely have equal numbers of positively charged and negatively charged groups that would pack well and not be bulky.

     (b)   Membrane proteins embedded in the membrane would likely have hydrophilic regions spanning the membrane with hydrophobic groups protruding from both sides. Protein transporters would be required for hydrophobic compounds that could not otherwise pass through the hydrophilic membrane interior.

     (c)   It may require hydrophilic solvents to solubilize these membranes and release embedded proteins prior to visualization by conventional means.

7-6.   (a)   Palmitate: 16 × 0.13 nm = 2.08 nm

     Laurate: 12 × 0.13 nm = 1.56 nm

     Arachidate: 20 × 0.13 nm = 2.6 nm

     (b)   The hydrophobic interior of a typical membrane is 4–5 nm.

     Two molecules of palmitate laid end to end: 4.16 nm

     Two molecules of laurate laid end to end: 3.12 nm

     Two molecules of arachidate laid end to end: 5.2 nm

Laurate (12C) molecules are too short to span the hydrophobic interior, whereas palmitate (16) and arachidate (20) are about the right length.

(c)   Each amino acid extends the long axis of the helix by about $0.56/3.6 = 0.156$ nm. To span a length of 4.16 nm (two palmitate molecules) will require $4.16/0.156 = 26.7$ or **about 27 amino acids.**

(d)   Seven transmembrane segments of about 26.7 amino acids each account for about 187 amino acids, which represent $187/248 = 0.752$, or **about 75% of the protein.** The remaining 61 (i.e., $248 \neq 187$) amino acids are present in the six hydrophilic loops that link the seven transmembrane segments together, so the hydrophilic loops must contain an average of about $61/6 = $ **10 amino acids.**

7-7.   Only (b) and (e) are unlikely. (b) is unlikely because longer-chain fatty acids decrease membrane fluidity, which is the opposite effect needed when the temperature has been lowered. (e) is unlikely because bacteria do not contain cholesterol under any conditions. On the other hand, (a) is likely because membrane fluidity is temperature dependent, and (c) and (d) are likely because unsaturated fatty acids will increase membrane fluidity.

7-8.   Membrane 1 has uniformly long and saturated fatty acids, so it has the *highest* transition temperature of the three (41°C). Double bonds are very disruptive of phospholipid packing in the membrane, so membrane 2 has the *lowest* transition temperature (–36°C). The shorter fatty acid "tails" of membrane 3 will lower the transition temperature noticeably but not drastically, so this membrane will have the *intermediate* transition temperature (23°C).

7-9.   (a)   Under these conditions, *Acholeplasma* cells can incorporate an appropriate combination of saturated and unsaturated fatty acids into their membranes to provide the cell with the optimum level of membrane fluidity.

(b)   Saturated fatty acids make a membrane less fluid. If only saturated fatty acids are available, the transition temperature of the membrane increases until the transition temperature is equal to the ambient temperature, at which point the membrane will gel.

(c)   The temperature of the culture could be raised to preserve membrane fluidity in spite of the prominence of saturated fatty acids in the membrane.

(d)   When a membrane gels, all cell functions that depend on the mobility of membrane proteins or lipids will be impaired or disrupted. Without the ability to transport solutes, detect and transmit signals, and carry out other membrane-dependent processes, a cell will not be able to survive.

(e)   Unsaturated fatty acids increase membrane fluidity, thus increasing the permeability of the membrane to ions and other solutes and making it impossible to maintain concentration gradients that are vital to life.

7-10.   (a)   To determine the structure of a protein by X-ray crystallography requires that the protein be isolated and crystallized. It is relatively straightforward to isolate membrane proteins but obtaining them in crystalline form has proven difficult to do with most integral membrane proteins. However, if an integral membrane protein can be isolated and sequenced, hydropathy analysis can be used to infer structural information.

(b)   A hydrophobic amino acid residue such as valine or isoleucine has a positive hydropathy index, whereas a hydrophilic residue such as aspartic acid or arginine has a negative hydropathy index.

(c)   Isoleucine is the most hydrophobic of the four amino acids, so it has the highest positive value: 3.1. Arginine is the most hydrophilic, so it has the most highly negative value: –7.5. Because of its hydroxyl group, serine is slightly more hydrophilic than alanine and therefore has the more negative of the two remaining values: –1.1. Alanine is slightly hydrophobic and has a positive value: 1.0.

(d)   A horizontal bar should be drawn over each of the seven peaks in the hydropathy plot, each of which corresponds to a transmembrane segment. The average transmembrane segment is about 20–30 amino acids long, which compares favorably with the value of 27 calculated in Problem 7-6c. The protein is in fact bacteriorhodopsin; see Figure 7-21b on p. 177 of the textbook.

7-11.   (a)   Some of the membrane proteins are associated with the outer phospholipid layer of the plasma membrane and protrude out from the membrane sufficiently to allow exposed tyrosine groups to be labeled by the LP reaction.

(b)   Some of the membrane proteins associated with the outer phospholipid layer of the plasma membrane are glycoproteins, the carbohydrate side chains of which are accessible to the GO and borohydride.

(c)   All the glycoproteins of the erythrocyte membranes are associated with the outer phospholipid layer of the membrane, and at least a portion of every carbohydrate side chain protrudes from the membrane surface sufficiently far to be labeled.

(d)   All the proteins associated with the outer phospholipid layer of the plasma membrane are glycoproteins; proteins bearing no carbohydrate side chains are, without exception, inaccessible to the labeling reagents.

(e)   All major membrane proteins protrude at least to some extent on one side of the membrane or the other; none is totally buried in the interior of the membrane.

7 12.   (a)   You would expect no labeling, because we already know from Problem 7-12c that all glycoproteins are associated with the outer layer and would therefore be on the interior of an inside-out vesicle.

(b)   You would expect to see labeling of all the proteins that were not labeled in Problem 7-12a, because we know from Problem 7-12e that almost all membrane proteins are accessible from one side of the membrane or the other.

(c)   You would conclude that at least some proteins extend all the way through the membrane and actually protrude sufficiently on both sides of the membrane to allow them to be labeled on either side.

(d)   Label membrane proteins of intact cells with one of the techniques described, then prepare inside-out vesicles and use the second labeling technique.

CHAPTER

8

Transport Across Membranes: Overcoming the Permeability Barrier

8-1. (a) F; the direction of movement depends on both the concentration gradient and the membrane potential, so facilitated diffusion can occur from a compartment of lower concentration to a compartment of higher concentration if the membrane potential is in the right direction and of sufficient magnitude.

(b) F; hydrolysis of high-energy phosphate bonds is a common means of driving active transport, but so is an ion gradient, as in sodium-driven cotransport of sugars and amino acids into cells.

(c) F; the K_{eq} for all uncharged solutes is 1.0; membrane permeability may affect the rate (or even the possibility) of movement, but it has no effect on the concentration ratio if and when equilibrium is reached.

(d) T

(e) T

(f) T

(g) F; carbon dioxide and bicarbonate ions move in opposite directions across the erythrocyte membrane in most cases.

(h) F; if the sodium-potassium pump is inhibited, the sodium gradient necessary for sodium-driven glucose uptake cannot be maintained and cotransport will decrease or even cease as the sodium gradient collapses.

8-2. (a) 3, 7 (c) 2, 3, 4, 6, 7 (e) 1 (g) 8

(b) 2, 4, 6 (d) 4, 5 (f) 5

8-3. (a) A, F (c) D (e) A (g) N (i) A, D, F

(b) D (d) A (f) D (h) D, F (j) A, F

8-4. Evidence that argues against the transverse carrier model: (a) Integral membrane proteins are embedded stably in the membrane and protrude from one or both sides based on their hydrophobic and hydrophilic regions; and (2) for a protein to traverse a membrane, movement of its hydrophilic region(s) through the hydrophobic interior of the membrane would be required, which would be highly endergonic and hence thermodynamically improbable.

8-5.   (a) $\Delta G_{inward} = RT \ln ([K^+]_{inside}/[K^+]_{outside}) = (1.987)(37 + 273) \ln 35$

$= 616 \ln (35) = (616)(3.55) = \textbf{+2190 cal/mol of potassium ions.}$

(b) $\Delta G_{inward} = RT \ln ([K^+]_{inside}/[K^+]_{outside}) + zFV_m$

$= +2187 \text{ cal/mol} + (1)(23,062)(-0.06 \text{ V})$

$= +2187 - 1384 = \textbf{+800 cal/mol of potassium ions.}$

(c)   Given the specified concentrations of ATP, ADP, and inorganic phosphate, we can calculate the free-energy change associated with the hydrolysis of one mole of ATP:

$\Delta G' = \Delta G^{\circ\prime} + RT \ln ([ADP][P_i]/[ATP]) = -7300 + 616 \ln (0.01/5)$

$= -7300 + 616 \ln 0.002 = -7300 + (616)(-6.215) = -7300 - 3828$

$= \textbf{--11,100 cal/mol of ATP molecules.}$

Mathematically, the 11,100 cal of energy released by the hydrolysis is theoretically enough to drive the inward pumping of about 13.9 (11,100/800) moles of potassium ions. However, any pumping mechanism transports an integral number of ions per ATP molecule, so the maximum possible number is **13 potassium ions pumped per molecule of ATP hydrolyzed.** No known pumping mechanism actually achieves this ratio, however. Even the sodium/ potassium pump responsible for most inward transport of potassium ions in animal cells moves only two potassium ions inward per mole of ATP hydrolyzed. (It is important to note, however, that this same pump moves sodium ions outward concomitantly, the energetics of which are considered in Problem 8-6c.)

8-6.   (a)   The sodium/potassium pump maintains gradients of both sodium and potassium ions using ATP as its energy source. Ion gradients can also be generated and maintained by exergonic electron transport.

(b)   An ion gradient is used to make ATP in both chemotrophic energy metabolism (Chapter 10) and photophosphorylation (Chapter 11).

(c)   Energy needed for outward pumping of sodium ions:

$\Delta G_{outward} = RT \ln ([Na^+]_{outside}/[Na^+]_{inside}) - zFV_m$

$= (1.987)(37 + 273) \ln (145/12) - (+1)(23,062)(-0.09 \text{ V})$

$= (616) \ln (12.1) - (-2076) = (616)(2.49) + 2076$

$= 1535 + 2076 = \textbf{3610 cal/mol of sodium ions.}$

Energy available from hydrolysis of ATP:

$$\Delta G' = \Delta G°' + RT \ln ([ADP][P_i]/[ATP]) = -7300 + 616 \ln (0.05/5)$$

$$= -7300 + 616 \ln 0.01 = 7300 + (616)(-4.605) = -7300 - 2837$$

$$= \textbf{-10,100 cal/mol of ATP hydrolyzed.}$$

Mathematically, the theoretical number of sodium ions that can be pumped outward per ATP hydrolyzed is 10,100/3610 = 2.80. However, any pumping mechanism transports an integral number of ions per ATP molecule, so ATP can be used to drive outward transport of Na^+ on a 2:1 basis but not on a 3:1 basis.

(d) Energy needed to synthesize ATP:

$$\Delta G' = \Delta G°' + RT \ln ([ATP]/[ADP][P_i]) = 7300 + (1.987)(25 + 273) \ln (5/0.05)$$

$$= 7300 + 592 \ln 100 = 7300 + (592)(4.605) = 7300 + 2726$$

$$= \textbf{10,030 cal/mol of ATP synthesized.}$$

Energy available from proton gradient:

$$\Delta G_{inward} = RT \ln ([H^+]_{inside}/[H^+]_{outside}) + zFV_m$$

$$= (1.987)(25 + 273) \ln (10^{-8}/10^{-7}) - (+1)(23,062)(-0.18 \text{ V})$$

$$= (592) \ln (0.1) - (-23,062)(0.18) = (592)(-2.303) + 4151$$

$$= -1363 - 4151 = \textbf{-5514 cal/mol of protons.}$$

The energy available from the proton gradient is not adequate to support ATP on a 1:1 basis. It is, however, adequate to support ATP synthesis on a 1:2 basis because the energy yield of two moles of protons ($2 \times -5514 = -11,028$ cal) is sufficiently exergonic (though only barely so) to provide the $-10,030$ cal needed for the synthesis of one mole of ATP.

8-7. (a) For ATP hydrolysis:

$$\Delta G = -7300 + RT \ln \frac{[ADP][P_i]}{ATP}$$

$$= -7300 + (1.987)(298) \ln (0.002)(0.001)/(0.020)$$

$$= -7300 + 592 \ln (0.0001) = -7300 + (592)(-9.210)$$

$$= -7300 + (-5452) = -12,800 \text{ cal/mol ATP.}$$

Available per sodium ion:

$$\Delta G = (-12,750 \text{ cal/mol ATP})(1 \text{ mol ATP}/3 \text{ mol Na}^+)$$

$$= -4250 \text{ cal/mol Na}^+$$

For outward sodium movement:

$$\Delta G_{outward} = RT \ln \frac{[\mathrm{Na^+}]_{outside}}{[\mathrm{Na^+}]_{inside}} - zEF$$

$$- 592 \ln \frac{0.15}{[\mathrm{Na^+}]_{inside}} - (+1)(-0.075)(23{,}062)$$

$$= 592 \ln \frac{0.15}{[\mathrm{Na^+}]_{inside}} + 1730 = +4250 \ \mathrm{cal/mol.}$$

$$\ln \frac{0.15}{[\mathrm{Na^+}]_{inside}} = (4250 - 1730)/592 = 2520/592 = 4.26$$

$$\frac{0.15}{[\mathrm{Na^+}]_{inside}} = e^{4.26} = 70.8$$

$$[\mathrm{Na^+}]_{inside} = 0.15/70.8 = 0.00212 = \textbf{2.12 m}\textbf{\textit{M}.}$$

(b)   If it were an uncharged molecule, the internal concentration could be reduced much more because the ATP-driven pumping would not have to "fight" the membrane potential. (The actual value for an uncharged molecule would be 0.11 mM.)

8-8.   (a)   The concentration gradient is $10^{-2}/10^{-7} = \textbf{10}^{\textbf{5}}.$

(b) $$\Delta G_{outward} = RT \ln \frac{[\mathrm{H^+}]_{inside}}{[\mathrm{H^+}]_{outside}}$$

$$= (1.987)(310) \ln (10^5) - (+1)(-0.07)(23{,}062)$$

$$= +7092 \ \mathrm{cal/mol} + 1613 \ \mathrm{cal/mol}$$

$$= +8705 \ \mathrm{cal/mol} = \textbf{+8.7 kcal/mol.}$$

(c) ΔG° for ATP hydrolysis is only –7.3 kcal/mol, but concentrations of ATP, ADP, and P$_i$ are usually such that ΔG for ATP hydrolysis in the cell is in the range of –10 to –14 kcal/mol, which would probably be adequate to drive hydrogen secretion on a 1:1 basis.

(d)   If the membrane potential (E) is to be just high enough to prevent the inward movement of protons when the outside-to-inside proton gradient is 10^5, then we can write

$$\Delta G_{inward} = RT \ln \frac{[H^+]_{inside}}{[H^+]_{outside}} + zEF = 0$$

or $E = (-RT/zF) \ln (10^{-5})$

$$= -(1.987)(310)/(+1)(23,062) \times \ln (10^{-5})$$

$$= (-0.0267)(-11.5) = +0.308 \text{ V} = \textbf{+308 mV.}$$

8-9. (a) The charged, or protonated, form of ammonia is the ammonium ion, NH_4^+.

(b) Under acidic conditions the protonated form, NH_4^+, will predominate because the high H^+ concentration will drive the equilibrium to the right:

$$NH_3 + H^+ \rightleftharpoons NH_4^+$$

(c) The inward movement of both ammonia and ammonium ion will be affected by their respective concentration gradients across the membrane. The membrane potential will affect the movement of ammonium ion only because it has a net charge, whereas ammonia does not.

(d) To say that the cell has a negative membrane potential means that the inner surface of the plasma membrane has a more negative charge than the outer surface. A positively charged solute such as NH_4^+ will therefore be drawn inward by the membrane potential, which means that uptake of NH_4^+ across the plasma membrane will require less energy than the uptake of NH_3—assuming the same concentration gradient for the two solutes, of course.

(e) The charged form of acetic acid is the ionized form, acetate:

$$CH_3-COOH \rightleftharpoons CH_3-COO^- + H^+$$

Uptake of both forms will be affected by their respective concentration gradients across the plasma membrane, but only the uptake of acetate ion will be affected by the membrane potential because it has a net charge, whereas acetic acid does not.

(f) Uptake of acetate will require more energy than the uptake of acetic acid because in this case the ion is negatively charged, which means that its uptake will be against the negative membrane potential.

8-10. (a) Figure 8-15 shows that 0.2 µmoles of ATP were hydrolyzed during an interval of 1 minute (beginning at 2 minutes and ending at 3 minutes) by 1.0 mg of protein present as added sarcoplasmic reticulum. The ATPase activity is therefore **0.2 µmoles/min per milligram of protein.**

(b) During the same 1-minute interval, all of the added calcium (0.4 µmoles) was taken up, as shown by the change in slope of the ATP hydrolysis rate at the end of the minute. The Ca^{2+}/ATP ratio is therefore 0.4 µmoles/0.2 µmoles = **2 calcium ions transported inward for each molecule of ATP that is hydrolyzed.**

(c)   When the ionophore is added at 4 minutes, the calcium that had been accumulated within the vesicles leaks out. The presence of calcium ions in the medium activates the calcium-dependent ATPase, so that ATP hydrolysis begins once again. The reaction will continue until all of the ATP has been hydrolyzed to ADP and phosphate.

8-11.   (a)   Such vesicles are free of endogenous energy sources and do not metabolize most substrates. However, you cannot be sure that a given transport system will be active in such vesicles, and it is possible that membrane proteins could alter their positions during vesicle formation.

(b) Na^+: outside; K^+: inside; ATP: outside.

(c)   ATP hydrolysis will continue at a high constant rate until the sodium and potassium concentration gradients across the membrane approach the maximum levels attainable with the given concentration of ATP, ADP, and P_i. The rate will then drop rapidly to the low baseline level of whatever ATP hydrolysis is necessary to replenish the gradient due to ion leakage across the membrane.

8-12.   (a)   No; sodium ion cotransport is required for active transport of glucose (by the Na^+/glucose symporter), but not for facilitated diffusion of glucose (by the glucose transporter).

(b)   Yes; cotransport of sodium ions drives the inward movement of amino acids and can only occur if sodium ions are actively pumped back out again.

(c)   Yes; potassium ions must be actively pumped into red blood cells, and this can occur only via a pump that couples the inward pumping of potassium ions to the outward pumping of sodium ions.

(d)   No; active uptake of sugars and amino acids in bacteria is driven by a proton gradient and is therefore not coupled to sodium cotransport as it is in animal cells.

9

Chemotrophic Energy Metabolism: Glycolysis and Fermentation

9-1. (a) F; energy is always *required* to break a covalent bond, including the phospho-anhydride bond that links the terminal phosphate group to the rest of the ATP molecule. ATP is a "high-energy" compound because its hydrolysis is exergonic, which means that more energy is released as the bonds between the –H and –OH groups of water are formed than is required to break the phosphoanhydride bond of ATP.

(b) T

(c) T

(d) F; the phosphate group doesn't "possess" any intrinsic energy of its own. The term "high-energy" applies only to the phosphoanhydride bond that links the phosphate group to the rest of the ATP molecule, and then only in the sense explained in the answer to part a.

(e) F; "low-energy bonds" require *more* energy to break, which is why less energy is released when such bonds are hydrolyzed.

(f) T

9-2. (a) The clue is in the word *zymase,* an early term for enzyme. The heat-labile fraction (zymase) contains the enzymes and the heat-stable fraction (cozymase) contains the coenzyme (NAD^+) necessary for fermentative activity. This observation is important in distinguishing enzymes from coenzymes and in understanding the need for both.

(b) Phosphate is required as substrate in reaction Gly-6, and the sequence cannot function without it. This observation is important in establishing the involvement of phosphate groups in energy metabolism.

(c) Based on the accumulation of a doubly phosphorylated sugar, we can deduce that iodoacetate must be an inhibitor of aldolase, the enzyme that splits fructose-1,6-bisphosphate into two trioses. In the presence of iodoacetate, aldolase is inactive and fructose-1,6-bisphosphate accumulates. This observation is important to establish fructose-1,6-bisphosphate as an intermediate in the pathway. (Iodoacetate is, in fact, a general inhibitor of Mg^{2+}-requiring enzymes, of which aldolase is an example.)

(d)   Fluoride ion is a potent inhibitor of enolase, the enzyme that catalyzes the removal of water from 2-phosphoglycerate to generate phosphoenolypyruvate in Reaction Gly-9. Addition of fluoride to the yeast extract caused the accumulation of two phosphorylated three-carbon acids, which could then be identified as 2- and 3-phosphoglycerate, thereby establishing the chemical nature of two further intermediates in the pathway.

9-3.   (a)   glycolysis is the first stage in aerobic energy metabolism for any cell that depends on glucose as its energy source, as brain cells do.

(b)   the oxidation that occurs at one step is balanced by the reduction that occurs when pyruvate is converted to lactate or ethanol and carbon dioxide.

(c)   whether an external electron acceptor is available, which in most cases means whether oxygen is available (i.e., whether the cell is functioning under aerobic conditions).

(d)   the pyruvate that is then reductively decarboxylated to generate the ethanol present in the beer and the carbon dioxide that causes the bread to rise.

(e)   the liver and heart muscle.

(f)   energy is always lost (as heat and entropy) whenever a reaction occurs, so glucose formation from lactate requires more energy than glucose catabolism releases.

9-4.   (a)   The glycosidic bond that links successive glucose units together in a polysaccharide has sufficient free energy of hydrolysis to allow it to be cleaved by phosphorolysis with the direct uptake of inorganic phosphate. The product is therefore a glucose molecule that is already phosphorylated (on carbon atom 1), which means that step Gly-1 of the glycolytic pathway is bypassed and the ATP that would otherwise be required there is saved.

(b)   In the intestine, sucrose is hydrolyzed to free glucose and fructose, each of which is then catabolized by the glycolytic pathway with the expected yield of 2 molecules of ATP per molecule of monosaccharide.

(c)   Because monosaccharide units are cleaved from a polysaccharide by phosphorolysis, we can suggest the same mechanism for bacterial sucrose catabolism. Phosphorolysis of a disaccharide will generate one phosphorylated monosaccharide that will yield three molecules of ATP by the glycolytic pathway (because step Gly-1 is bypassed and one less ATP molecule is needed) and one free monosaccharide that will require step Gly-1 and will therefore yield two molecules of ATP. In fact, the sucrose phosphorylase reaction generates glucose-1-phosphate and free fructose, and the glycolytic ATP yield is therefore 5 molecules of ATP per molecule of sucrose, or 2.5 ATP molecules per monosaccharide.

(d)   Raffinose has three monosaccharides linked together by two glycosidic bonds. Phosphorolysis of these bonds generates two phosphorylated hexoses (yield: 3 ATP each) and one free hexose (yield: 2 ATP). The average ATP yield per monosaccharide is therefore $(3 + 3 + 2)/3 = $ **2.67.**

9-5. (a) $$\Delta G' = \Delta G^{\circ\prime} + RT \ln \frac{[\text{glucose-6-phosphate}]}{[\text{glucose}]\,[P_i]}$$

$$= +3300 + (1.987)(298) \ln \frac{0.08 \times 10^{-3}}{[\text{glucose}](1.0 \times 10^{-3})}$$

$$= +3300 + 592 \ln \frac{0.08}{[\text{glucose}]}$$

$$= +3300 + 592 \ln (0.08) - 592 \ln [\text{glucose}]$$

$$= +3300 - 1495 - 592 \ln [\text{glucose}]$$

At equilibrium, $\Delta G' = 0$, so $\Delta G' = 1805 - 592 \ln [\text{glucose}] = 0$.

Because $\ln [\text{glucose}] = \dfrac{1805}{592} = 3.049$, $[\text{glucose}] = e^{3.049} = $ **21 *M*!**

This means that a glucose concentration of 21 *M* would be required just to bring the reaction to equilibrium; anything over this would render the reaction spontaneous in the direction of phosphorylation. This is impossible; even 2 *M* glucose would be a thick syrup!

(b) Glucose + ATP $\longrightarrow$ glucose-6-phosphate + ADP

$\Delta G^{\circ\prime} = +3.3$ kcal/mol $- 7.3$ kcal/mol = **−4.0 kcal/mol.**

(c) $$\Delta G' = \Delta G^{\circ\prime} + RT \ln \frac{[\text{glucose-6-phosphate}]\,[\text{ADP}]}{[\text{glucose}]\,[\text{ATP}]}$$

$$= -4000 + (1.987)(298) \ln \frac{(0.08 \times 10^{-3})(0.15 \times 10^{-3})}{[\text{glucose}]\,(1.8 \times 10^{-3})}$$

$$= -4000 + 592 \ln \frac{(6.67 \times 10^{-6})}{[\text{glucose}]}$$

$$= -4000 + 592 \ln (6.67 \times 10^{-6}) - 592 \ln [\text{glucose}]$$

$$= -4000 - 7055 - 592 \ln [\text{glucose}] = 11{,}055 - 592 \ln [\text{glucose}]$$

At equilibrium, $\Delta G' = 0$, so $\Delta G' = -11{,}055 - 592 \ln [\text{glucose}] = 0$. Because $\ln [\text{glucose}] = -11{,}055/592 = -18.67$, $[\text{glucose}] = e^{-18.67} = $ **7.76 × 10⁻⁹ *M*!** This means that a glucose concentration of 7.76×10^{-9} *M* would bring the reaction to equilibrium; any glucose concentration higher than this will render the reaction spontaneous in the direction of phosphorylation. This is physiologically very reasonable, because glucose phosphorylation is thermodynamically feasible as long as the glucose concentration remains more than 0.01 µ*M*.

(d)   From 2.1×10^1 to 7.76×10^{-9} is more than nine orders of magnitude!

(e) $\Delta G' = \Delta G^{\circ\prime} + RT \ln \dfrac{[\text{glucose-6-phosphate}][\text{ADP}]}{[\text{glucose}][\text{ATP}]}$

$= -4000 + 592 \ln \dfrac{(0.08 \times 10^{-3})(0.15 \times 10^{-3})}{(5.0 \times 10^{-3})(1.8 \times 10^{-3})}$

$= -4000 + 592(-6.623)$

$= -4000 - 3920 = -7920^{\text{cal/mol}} = \textbf{--7.9 kcal/mol.}$

9-6.   (a)   Ethanol catabolism in the body begins with its oxidation (dehydrogenation), with NAD^+ as the electron acceptor. The more ethanol that is consumed, the greater the demand for NAD^+ and the more serious is the reduction in NAD^+ concentration. This means, in turn, that the supply of NAD^+ may be inadequate for aerobic respiration of glucose.

(b)   Acetaldehyde is the immediate product of ethanol oxidation:

Ethanol + NAD^+ → acetaldehyde + NADH + H^+

(c)   Methanol and ethanol are both substrates of the enzyme alcohol dehydrogenase and therefore compete for the active site. The body is flooded with a large amount of ethanol to provide an effective competitor of methanol, thereby minimizing the production of formaldehyde and lessening the danger that the patient will be "pickled."

9-7.   (a)   Propionate differs from pyruvate in that its middle carbon atom is at the hydrocarbon level rather than at the carbonyl level. To reduce pyruvate to propionate would therefore require two molecules of NADH, assuming an appropriate sequence of reactions exists. But the stoichiometry of glycolysis provides only one molecule of NADH per molecule of pyruvate, not two. Therefore, all of the pyruvate cannot be reduced to propionate.

(b)   If 50% of the pyruvate molecules are reduced to propionate and the remaining 50% are left as pyruvate, the stoichiometry will come out right. The overall reaction would then be

$C_6H_{12}O_6 \longrightarrow CH_3\text{--}CO\text{--}COO^- + CH_3\text{--}CH_2\text{--}COO^- + H_2O + 2\ H^+$

Glucose   Pyruvate   Propionate

(c)   The pyruvate must be further metabolized by decarboxylation.

9-8.   (a)   A reaction sequence that is thermodynamically feasible (exergonic) in one direction will not function in the other direction by simple reversal of each of the reactions because it will be endergonic in that direction under the same conditions. (Theoretically, one can imagine conditions in which a sufficiently high concentration of product(s) and a sufficiently low concentration of reactant(s) might drive a reaction sequence in the opposite direction, but the thermodynamic driving force—that is, the $\Delta G'$ for the reaction sequence in the forward direction—is usually sufficiently great for most metabolic pathways that it is virtually impossible to reverse such pathways simply by changes in reactant and product concentrations.)

(b)   2 pyruvate + 4 ATP + 2 GTP + 6 H_2O + 2 NADH + 2 H^+ →

glucose + 4 ADP + 2 GDP + 6 P_i + 2 NAD^+

(c)   The four additional phosphoanhydride bonds provide the extra energy needed to ensure that the pathway is driven strongly enough in the gluconeogenic direction so that it is essentially irreversible in that direction.

(d)   The $\Delta G'$ for the glycolytic pathway under typical cellular conditions is about –20 kcal/mol and therefore + 20 kcal/mol in the opposite direction. Because four additional phosphoanhydride bonds are hydrolyzed to drive the pathway in the opposite direction and each of those bonds has a $\Delta G'$ of about –10 kcal/mol, the net driving force in the gluconeogenic direction is

$$\Delta G' = 20 + 4(-10) = \textbf{–20 kcal/mol.}$$

(e)   The key glycolytic and gluconeogenic enzymes shown in Figure 9-12 on p. 244 of the textbook are subject to regulation by a variety of factors, including the AMP, ADP, ATP, acetyl CoA, and F2,6BP status of the cell. Changes in the concentrations of these intermediates either activate or inhibit the regulatory enzymes, thereby effectively turning the pathway "on" in one direction and "off" in the other.

9-9.   (a)   The discovery that glycolytic enzymes are compartmentalized in trypanosomes came about as a result of differential centrifugation, a technique that causes most organelles to sediment to the bottom of a centrifuge tube in response to centrifugal force while molecules, ions, and smaller organelles such as ribosomes remain in the supernatant. In an experiment in which glycolytic enzymes were expected to remain in the supernatant, seven out of the ten enzymes did not, but were recovered in the pellet instead, indicating an organellar location.

(b)   With the glycolytic enzymes compartmentalized in this way, the enzymes and intermediates in the pathway can be maintained in relatively high concentrations because the organellar volume is small compared to the volume of the cell.

(c)   The first seven steps of the glycolytic pathway begin with glucose and lead to the formation of 3-phosphoglycerate. The glycosomal membrane must therefore have transport proteins for glucose (which moves inward), 3-phosphoglycerate (which moves outward), and inorganic phosphate (which must move inward to balance the loss of phosphate as 3-phosphoglycerate moves outward). (Bonus question: Can you suggest a means whereby the outward movement of 3-phosphoglycerate can be coupled to the inward movement of inorganic phosphate?)

(d)   By biochemical definition, a peroxisome is capable of carrying out the generation and degradation of hydrogen peroxide (H_2O_2) and always possesses at least one H_2O_2-generating enzyme (an oxidase) and one H_2O_2-degrading enzyme (catalase).

9-10.   (a)   The ATP yield per unit of glucose catabolized is much lower under anaerobic conditions (2 ATP/glucose) than under aerobic conditions (36–38 ATP/glucose). A yeast cell functioning under anaerobic conditions must therefore consume 18 to 19 times as much glucose per unit time as under aerobic conditions in order to sustain the same rate of ATP generation.

(b)   Although the glycolytic pathway responsible for glucose fermentation by yeast cells begins and ends with an unphosphorylated molecule (glucose and ethanol, respectively), almost all of the intermediates in the process are phosphorylated compounds. The fermentation process therefore requires inorganic phosphate, which is taken up in step Gly-6 and used to generate ATP from ADP in step Gly-7, and will be stimulated by the addition of inorganic phosphate if phosphate is a limiting reagent in the culture medium.

(c)   The ATP needed for the rapid but short-term movements of an alligator is generated by glycolysis at the expense of muscle glycogen. The long period of recovery following such movements is needed for the replenishment of muscle glycogen stores.

(d)   The oxidation of glyceraldehyde-3-phosphate that occurs at step Gly-6 involves the oxidation of a carbonyl group to a carboxylic acid group, which is a highly exergonic reaction. The reduction of pyruvate to lactate is endergonic but involves the reduction of a carbonyl group to a hydroxyl group, a reaction that requires less energy than is released by the oxidation of a carbonyl group to a carboxylic acid group. (Note, then, that anaerobic life is possible only because of the difference in energy levels between a carboxyl group and a hydroxyl group!)

9-11.   (a)   It allows substrate oxidation to proceed without concomitant ATP generation, releasing this step in the glycolytic pathway from its normal sensitivity to, and regulation by, the availability of ADP and P_i.

(b)   Use of arsenate instead of phosphate at step Gly-6 results in spontaneous hydrolysis of the arseno intermediate without conservation of the energy of the bond as ATP. This results in two molecules less of ATP per molecule of glucose, so the energy yield under anaerobic conditions is zero, and the arsenate is therefore fatal.

(c)   Any reaction involving the direct uptake of inorganic phosphate that leads to the generation of a high-energy phosphate bond and the formation of ATP may be subject to uncoupling in this way, provided only that the enzyme will accept arsenate in place of phosphate at its active site.

9-12. (a)

6-Phosphogluconolactone   Glucose 6-phosphate

– H₂O

Gluconolactonase

→ H⁺

(b)

6-Phosphogluconate

H₂O ←   6-Phosphogluconate dehydratase

(c)

2-Keto-3-deoxy-6-phosphogluconate (KDPG)

KDPG Aldolase →

Pyruvate

Glyceraldehyde 3-phosphate

(d) The ATP yield per glucose will be only one instead of two. Only one molecule of ATP is required as an input (to form the glucose-6-phosphate), but only one molecule of glyceraldehyde-3-phosphate is produced, generating two molecules of ATP as it is oxidized to pyruvate.

(e) In bacteria using the Entner-Doudoroff pathway, the amount of lactic acid production in the presence of arsenate would likely not change. Arsenate blocks only the production of ATP (Gly-7) and not the production of NADH (Gly-6).

9-13. (a)   Fructose-2,6-bisphosphate (F2,6BP) is an allosteric activator of phosphofructo-kinase (PFK). The enzyme shows normal Michaelis-Menten kinetics in the presence of F2,6BP (red line) but not in its absence (black line). In fact, the apparent K_m of the enzyme for fructose-6-phosphate is at least five times higher in the absence of F2,6BP than in its absence. F2,6BP, in other words, is required for normal functioning of PFK. (The sigmoidal dependence of reaction rate of substrate concentration in the absence of F2,6BP is characteristic of an allosterically regulated enzyme when functioning in the absence of its allosteric effector.)

(b)   ATP is an allosteric inhibitor of PFK. The enzyme shows normal Michaelis-Menten kinetics when the ATP concentration is low (black line) but not when the ATP concentration is high (red line). In fact, the apparent K_m of the enzyme for fructose-6-phosphate is at least five times higher in the presence of a high concentration ATP than it is in when the ATP concentration is low. In other words, the ATP concentration must be kept low for PFK to function normally. (Note, however, that ATP is not only an allosteric effector of PFK but also one of its substrates, so normal functioning of the enzyme requires an ATP concentration that is low enough to avoid allosteric inhibition but high enough to sustain catalytic activity at the active site.)

(c)   In Figure 9-14a, we must assume that the ATP concentration is high enough to sustain catalytic activity (i.e., significantly greater than the K_m of the enzyme for ATP) but still low enough to avoid allosteric inhibition by ATP. In Figure 9-14b, we must assume that the F2,6BP concentration is high enough (0.13 mM, for example) to ensure allosteric activation of the enzyme.

CHAPTER
10
Chemotrophic Energy Metabolism: Aerobic Respiration

10-1. (a) M   (d) IM   (g) IM   (j) IS   (m) IM
       (b) IM   (e) M   (h) NO   (k) IM
       (c) IS   (f) OM   (i) IM   (l) IS

10-2. (a) C   (d) PM   (g) NO   (j) EX   (m) NO
       (b) PM   (e) C   (h) C   (k) PM
       (c) C   (f) PM   (i) PM   (l) NO

10-3. (a) F; the orderly flow of carbon through the TCA cycle is possible because most
            (all but one) of the enzymes of the cycle are present in soluble form in the
            mitochondrial matrix, where the substrate(s) of each enzyme can collide with,
            and bind specifically to, the active site of the enzyme.

       (b) T

       (c) F; respiration is an aerobic process in many organisms because oxygen is the
            single most common electron acceptor for reoxidation of reduced coenzymes.
            However, other electron acceptors such as S, H^+, or Fe^{3+} can be used in the
            absence of oxygen (anaerobic respiration).

       (d) T

       (e) T

       (f) F; fatty acids with an odd number of carbons are degraded by β-oxidation until
            three carbons of the original fatty acid remain. The remaining propionyl CoA
            molecule is then degraded in the TCA cycle after conversion to succinyl CoA.

10-4. (a) In; 2   (e) No flux   (j) No flux
       (b) In; 6   (f) No flux   (j) Out; 42
       (c) Out; 36   (g) No flux   (k) In; 2 pairs per glucose
       (d) In; 36   (h) No flux   (l) In and out; no net flux

10-5. (a) The pathway from citrate to α-ketoglutarate is as shown in Figure S10-1.

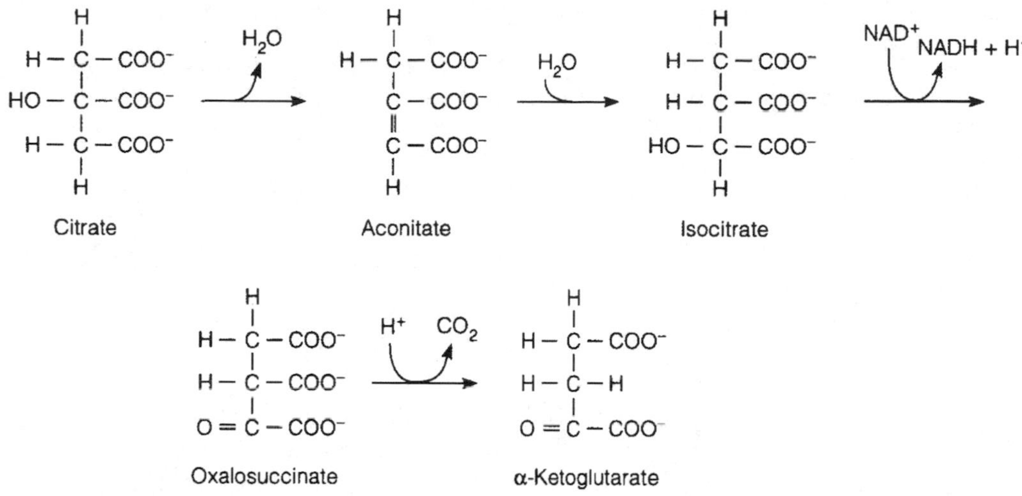

Figure S10-1   The Metabolic Path from Citrate to α-Ketoglutarate. The conversion of citrate into α-ketoglutarate proceeds in a four-step sequence involving dehydration, hydration, oxidation, and decarboxylation reactions. See Problem 10-5.

(b) The pathway from pyruvate and alanine to glutamate is as follows:

Pyruvate + CO_2 + ATP + $H_2O \longrightarrow$ oxaloacetate + ADP + P_i

Pyruvate + CoA–SH + $NAD^+ \longrightarrow$ acetyl CoA + CO_2 + NADH + H^+

Acetyl CoA + oxaloacetate + $H_2O \longrightarrow$ citrate + CoA–SH

Citrate $\longrightarrow$ isocitrate

Isocitrate + $NAD^+ \longrightarrow \alpha$-ketoglutarate + CO_2 + NADH + H^+

α-Ketoglutarate + alanine $\longrightarrow$ glutamate + pyruvate

---

Pyruvate + alanine + ATP + 2 H_2O + 2 $NAD^+ \longrightarrow$

glutamate + CO_2 + ADP + P_i + 2 NADH + 2 H^+

(c) By following the middle carbon atom of pyruvate in Figure 10-8 on p. 261 of the textbook, you should be able to convince yourself that it becomes the carboxyl carbon of acetate and hence the newly added ("upper") carboxyl group in citrate.

10-6. (a) Yes, isocitrate can pass electrons exergonically to NAD^+ under standard conditions because the α-ketoglutarate/isocitrate redox pair has a more negative E_0' (–0.38V) than does the NAD^+/NADH redox pair (–0.32V), which means that the reduced form of the isocitrate/α-ketoglutarate redox pair (i.e., isocitrate) will spontaneously reduce the oxidized form of the NAD^+/NADH redox pair (i.e., NAD^+).

(b)   The E_0' value for the reduction of NAD^+ by isocitrate is calculated as

$$\Delta E_0' = E_{0,acceptor}' - E_{0,donor}' = -0.32V - (-0.38V) = -0.32 + 0.38 = \textbf{+0.06V.}$$

The $\Delta E_0'$ value is positive, so the transfer of electrons from isocitrate to NAD^+ is thermodynamically spontaneous under standard conditions, as predicted in part a.

(c)   The $\Delta G^{o'}$ value for the reduction of NAD^+ by isocitrate is calculated as

$$\Delta G^{o'} = -nF\Delta E_0' = -1(23,062)(+0.06) = -1394 \text{ cal/mol} = \textbf{-1.38 kcal/mol.}$$

This is relevant to aerobic energy metabolism because the oxidation of isocitrate to α-ketoglutarate is one of the reactions of the TCA cycle (Reaction TCA-3).

(d)   Lactate *cannot* pass electrons exergonically to NAD^+ under standard conditions because the pyruvate/lactate redox pair has a less negative E_0' (−0.19V) than does the NAD^+/NADH redox pair (−0.32V), which means that the reduced form of the NAD^+/NADH redox pair (i.e., NADH) will spontaneously reduce the oxidized form of the pyruvate/lactate redox pair (i.e., pyruvate). Thus, the reaction will be spontaneous in the direction of pyruvate reduction to lactate, not lactate oxidation to pyruvate.

The $\Delta E_0'$ value for the reduction of lactate to pyruvate by NAD^+ is calculated as

$$E_0' = E_{0,acceptor}' - E_{0,donor}' = -0.32V - (-0.19V) = -0.32 + 0.19 = \textbf{-0.13V.}$$

The $\Delta E_0'$ value is negative, so the transfer of electrons from lactate to NAD^+ *is not* thermodynamically spontaneous under standard conditions, as predicted above.

The $\Delta G^{o'}$ value for the reduction of NAD^+ by isocitrate is calculated as

$$\Delta G^{o'} = -nF\Delta E_0' = -1(23,062)(-0.13) = +2998 \text{ cal/mol} = \textbf{+3.00 kcal/mol.}$$

These calculations are relevant to aerobic energy metabolism because in at least some types of cells (muscle cells, for example), lactate accumulates during periods of hypoxia (low oxygen concentration) and must be reoxidized to pyruvate when the oxygen concentration rises again. The above calculations tell us that this reoxidation would not be possible under standard conditions, but under cellular conditions, the concentration ratios of lactate to pyruvate and of NAD^+ to NADH must be sufficiently high to render the reaction exergonic.

(e)   By the same reasoning as in part d succinate *cannot* pass electrons exergonically to NAD^+ under standard conditions because the fumarate/succinate redox pair has a much less negative E_0' (−0.03V) than does the NAD^+/NADH redox pair (−0.32V), which means that the reduced form of the NAD^+/NADH redox pair (i.e., NADH) will spontaneously reduce the oxidized form of the fumarate/succinate redox pair (i.e., fumarate). Thus, the reaction will be spontaneous in the direction of fumarate reduction to succinate, not succinate oxidation to fumarate. The relevant calculations for $\Delta E_0'$ and $\Delta G^{o'}$ for succinate oxidation by NAD^+ are

$$\Delta E_0' = E_{0,acceptor}' - E_{0,donor}' = -0.32V - (-0.03V) = -0.32 + 0.03 = -0.29V.$$

$$\Delta G^{o\prime} = -nF\Delta E_0' = -1(23,062)(-0.29) = +6668 \text{ cal/mol} = +6.67 \text{ kcal/mol.}$$

The highly positive value for $\Delta G^{o\prime}$ makes it virtually certain that succinate oxidation with NAD^+ as the electron acceptor is impossible under cellular conditions. It is highly unlikely that the concentration ratios of the products and reactants could overcome so highly positive at $\Delta G^{o\prime}$ value. Thus, NAD^+ is *not* likely to serve as the electron acceptor for the succinate dehydrogenase reaction of the TCA cycle.

(f)   By the same reasoning as in part a, succinate *is* able to pass electrons exergonically to coenzyme Q under standard conditions because the fumarate/succinate redox pair has a more negative E_0' (−0.03V) than does the CoQ/CQH₂ redox pair (+0.04V). Thus, the reaction will be spontaneous in the direction of succinate oxidation under standard conditions. The relevant calculations for E_0' and $\Delta G^{o\prime}$ for succinate oxidation by CoQ are

$$E_0' = E_{0,acceptor}' - E_{0,donor}' = +0.04V - (-0.03V) = +0.04 + 0.03 = +0.07V.$$

$$\Delta G^{o\prime} = -nF\Delta E_0' = -1(23,062)(+0.07) = -1614 \text{ cal/mol} = -1.61 \text{ kcal/mol.}$$

These calculations mean that the transfer of electrons from succinate to CoQ will be thermodynamically spontaneous under standard conditions. It makes sense to regard coenzyme Q as the electron acceptor even though FAD is shown as the immediate electron acceptor in Figure 10-8 because the FAD involved here is tightly bound to the succinate dehydrogenase complex and passes its electrons to coenzyme Q, making the latter the eventual electron acceptor for this reaction.

10-7.   (a)   For calculation of the maximum ATP yield for an aerobic prokaryote, see Table S10-1 (but ignore numbers in parentheses; they are the values for the eukaryotic cell described in part b). The maximum ATP yield is obtained by summing the bottom line across the table: ATP yield = 8 + 6 + 24 = **38 ATP/glucose.**

Table S10-1   Calculation of Maximum ATP Yield from Aerobic Oxidation of Glucose

Stage of Respiration	Glycolysis (glucose → 2 pyruvate)	Pyruvate Oxidation (2 pyruvate → 2 acetyl CoA)	TCA Cycle (2 turns)
Yield of CO_2	0	2	4
Yield of NADH	2	2	6
ATP per NADH	3 (2)	3	3
Yield of $FADH_2$	0	0	2
ATP per $FADH_2$			2
ATP from substrate-level phosphorylation	2	0	2
ATP from oxidative phosphorylation	6 (4)	6	22
MAXIMUM ATP YIELD	8 (6)	6	24

(b)   For a eukaryotic cell that uses the glycerol phosphate shuttle, substitute the parenthetical values shown in the "Glycolysis" column; because the glycerol phosphate shuttle transfers electrons from NADH in the cytosol to FAD in the mitochondrion, the ATP yield per cytoplasmically generated ATP is 2 instead of 3, and the ATP yield is therefore decreased by 2: ATP yield = 6 + 6 + 24 = **36 ATP/glucose.**

10-8. (a)   High levels of ADP mean low levels of ATP, so it is to the advantage of the cell to activate the metabolic pathway responsible for coenzyme reduction, which can in turn give rise to ATP synthesis by electron transport.

(b)   High NADH levels mean adequate reduced coenzyme for the generation of more ATP, so it makes sense to shut down the catabolic machinery of the cell.

(c)   High ATP levels indicate adequate energy supply, so it makes sense that the enzyme responsible for providing the TCA cycle with more acetyl CoA substrate is shut down.

(d)   High citrate levels are indicative of a sufficient supply of acetyl CoA, so it is reasonable that the key regulatory enzyme of the pathway leading to pyruvate and acetyl CoA is decreased in activity.

(e)   High levels of NADH mean adequate reduced coenzyme for the generation of ATP, so it makes sense to convert PDH to the inactive (phosphorylated) form, which is what PDH kinase does.

(f)   High levels of succinyl CoA signal adequate levels of TCA-cycle intermediates, so it seems reasonable to shut down further TCA cycle activity.

10-9. (a)   Fluorocitrate has been characterized as the actual poison in the tissues of the animal, and one of the most pronounced of its effects is a buildup of at least one of the TCA cycle intermediates.

(b)   The suspected blockage point is the conversion of citrate to isocitrate (inhibition of the enzyme aconitase), because (1) it is specifically citrate that accumulates in the tissues of poisoned animals and (2) fluorocitrate is an analogue of citrate and might be expected to bind to and thereby to block the active site of the enzyme that metabolizes citrate in the TCA cycle.

(c)   Fluoroacetate is probably activated to fluoroacetyl CoA and condensed onto oxaloacetate by citrate synthase to generate fluorocitrate.

(d)   The ingested compound is itself not toxic, but it is converted into a lethal metabolite in vivo (that is, a lethal compound is synthesized in vivo from a nonlethal precursor).

10-10. (a)   Degradation of the 16-carbon palmitate molecule will require 7 cycles of β-oxidation, producing 7 $FADH_2$, 7 NADH, and 8 acetyl CoA.

(b)   Degradation of the 18-carbon stearate molecule will require 8 cycles of β-oxidation, producing 8 $FADH_2$, 8 NADH, and 9 acetyl CoA.

(c)   For a fatty acid with a total of n carbons, the equation would be:

$CH_3–(CH_2)_{(n-2)}–COO^- + ATP \rightarrow n/2$ acetyl CoA $+ ((n/2) – 1)$ FADH$_2$ + $((n/2) – 1)$ NADH

(d)   Palmitate: Each of the 8 acetyl CoA molecules will produce 12 ATP in the TCA cycle (3 NADH + 1 FADH$_2$ + 1 ATP per acetyl CoA = 9 + 2 + 1 = 12 ATP per acetyl CoA), yielding 96 ATP. The 7 FADH$_2$ will yield 14 ATP, and the 7 NADH will yield 21 ATP, for a total of 131 ATP. Because 1 ATP was consumed in the initial activation step, the grand total will be 130 ATP produced following the complete oxidation of one molecule of palmitate to 16 molecules of CO$_2$. Complete oxidation of stearate to 18 molecules of CO$_2$ will yield 147 molecules of ATP.

(e)   For a fatty acid with a total of n carbons, the equation would be:

Total ATP $= [n/2$ acetyl CoA $\times 12$ ATP/acetyl CoA$] + [((n/2) – 1)$ FADH$_2$
$\times 2$ ATP/FADH$_2] + [((n/2) – 1)$ NADH $\times 3$ ATP/NADH$] – 1$
$= 6n + [n – 2] + [(3n/2) – 3] – 1$
$= 8.5n – 6$

10-11. (a)   DHAP + NADH$_{cytosol}$ + H$^+$ $\xrightarrow[\text{in the cytosol}]{}$ Gly-3-P + NAD$^+_{cytosol}$

FAD$_{mitochondrion}$ + Gly-3-P $\xrightarrow[\text{in the inner membrane}]{}$ DHAP + FADH$_{2,mitochondrion}$

(b)   NADH$_{cytosol}$ + H$^+$ + FAD$_{mitochondrion}$ $\rightarrow$ NAD$^+$ cytosol + FADH$_{2,mitochondrion}$

To calculate the difference in standard reduction potentials, use Equation 10-9 on p. 272 of the text and the appropriate data from Table 10-2 on the same page.

$\Delta E_0' = \Delta E_0'_{,acceptor} – \Delta E_0'_{,donor} = –0.18 – (–0.32) = $ **+0.14 V**

To calculate the corresponding standard free-energy change, use Equation 10-11 on p. 273 of the textbook:

$\Delta G^{\circ\prime} = –nF\Delta E_0' = –2(23,062)(+0.14) = –6460$ cal/mol $= $ **–6.46 kcal/mol.**

The inward movement of electrons is feasible, given the highly negative $\Delta G^{\circ\prime}$ value.

(c)   FADH$_{2,mitochondrion}$ + COQ$_{mitochondrion}$ $\rightarrow$ FAD$_{mitochondrion}$ + CoQH$_{2,mitochondrion}$

$\Delta E_0' = \Delta E_0'_{,acceptor} – \Delta E_0'_{,donor} = +0.04 – (–0.18) = $ **+0.22 V**

$\Delta G^{\circ\prime} = –nF\Delta E_0' = –2(23,062)(+0.22) = –10,150$ cal/mol $= $ **–10.1 kcal/mol.**

This transfer is also thermodynamically feasible under standard conditions.

(d)   NADH$_{cytosol}$ + H$^+$ + CoQ$_{mitochondrion}$ $\longrightarrow$ NAD$^+_{cytosol}$ + CoQH$_{2,mitochondrion}$

Because this reaction is the sum of the reactions in parts b and c, the values for $\Delta E_0'$ and $\Delta G^{\circ\prime}$ can be obtained as the sums of the values for the two component reactions:

$\Delta E_0' = +0.14 + 0.22$ V $= $ **+0.36 V**

$\Delta G^{\circ\prime} = 6460 – 10,150 = –16,600$ cal/mol $= $ **–16.1 kcal/mol.**

Given that both component reactions are exergonic under standard conditions, it is not surprising that the overall reaction is exergonic—*highly* exergonic, in fact.

(e) $\Delta G' = \Delta G^{\circ\prime} + RT \ln ([NAD^+] [CoQH_2]/[NADH] [CoQ])$

$= -16,600 + (1.987) (25 + 273) \ln (2.0/5.0)$

$= -16,600 + 592 \ln (0.4) = -16,600 + (592) (-0.916)$

$= -16,600 - 542 = -16,060 \text{ cal/mol} = \textbf{-16.1 kcal/mol.}$

(f)   No, it is not affected by the concentration of the intermediate in the process because free energy is a thermodynamic parameter and is therefore a measure of the difference in energy content between the reactants and products only.

10-12. (a)   Given that ATP synthesis does not occur, the energy is lost as heat. For a newborn baby, the heat generated in this way may be critical to maintenance of body temperature.

(b)   One would expect to find more thermogenin in a hiberating bear because the need for additional heat is clearly more critical during hibernation, when the external temperature is likely to be colder and bodily activity much less than in the case of a physically active bear.

(c)   The localization of thermogenin to the inner mitochondrial membrane and its mode of action as an uncoupler of electron transport make it likely that thermo-genin is a proton translocator that allows electrons to move exergonically into the matrix of the mitochondrion just as F_o does, but without concomitant ATP generation. To test this hypothesis, one could prepare vesicles consisting of phospholipid bilayers with and without thermogenin. The vesicles could be prepared in a mildly alkaline solution (pH 8.0, for example), then transferred to an acidic solution (pH 5.0, for example). If thermogenin is a proton translocator, protons should enter the vesicles containing this protein, such that the pH within the vesicle should quickly equilibrate with the external pH (i.e., pH 5), whereas the pH within the vesicles prepared without thermogenin should remain at the original pH (i.e., pH 8).

(d)   If all of the mitochondria were equipped with uncoupling protein, the mitochondria would not be able to produce much ATP. The organism would therefore have to depend on glycolysis for its ATP synthesis, which would almost certainly not be adequate for an organism that usually depends on mitochondrial ATP synthesis for most of its energy. Furthermore, all of the energy that would normally have gone into ATP synthesis would be liberated as heat so we would end up with an energy-deprived, overheated mammal!

10-13. (a)   The electron donor in this system is β-hydroxybutyrate, which can be oxidized to β-ketobutyrate. The electron acceptor is oxidized cytochrome c, which can be reduced by taking up a single electron. The most likely pathway of electron transport is from β-hydroxybutyrate via NAD^+, complex I, and complex III to cytochrome c.

(b)   Assuming the pathway specified in part a, the ATP yield is likely to be 2 moles of ATP per mole of β-hydroxybutyrate oxidized, one each by complexes I and III.

The reaction in this system is as follows:

$$\beta\text{-hydroxybutyrate} + 2 \text{ cyt } c_{\text{oxidized}} + 2\text{ADP} + 2\text{ P}_i \rightarrow$$

$$\beta\text{-ketobutyrate} + 2 \text{ cyt } c_{\text{reduced}} + 2\text{H}^+ + 2\text{ATP} + 2\text{H}_2\text{O}$$

(c) Cyanide is a potent inhibitor of cytochrome c oxidase. It is present to ensure that the electrons picked up by cytochrome c are not passed on down the system to O_2, which is the acceptor when the whole electron transport system is operative. In the absence of cyanide, the electrons from β-hydroxybutyrate would be transferred all the way to oxygen, which would defeat the purpose of the experiment.

(d) The TCA cycle would not be active because no substrate for the TCA cycle (i.e., no pyruvate, acetyl CoA, or fatty acid) is present in the assay system.

(e) It is important that the β-ketobutyrate generated by oxidation of β-hydroxybutyrate cannot be further metabolized, lest other oxidative reactions occur that would generate alternative substrates for the electron transport system. If lactate had been used instead—and if isolated mitochondria can oxidize lactate as they can oxidize β-hydroxybutyrate—pyruvate would be formed, which would activate the TCA cycle, thereby generating additional substrates for the electron transport system (NADH and FADH$_2$) and rendering the experiment useless for its intended purpose.

11

Phototropic Energy
Metabolism: Photosynthesis

11-1. (a)   F   (c)   F   (e)   T

      (b)   F   (d)   F

11-2. (a)   You would expect noncyclic electron flow to decrease as electron flow from PSI to $NADP^+$ is diverted into cyclic electron flow. The researchers found that noncyclic electron flow in *hcef* mutant plants was decreased by 80% compared to normal plants.

      (b)   You would expect the light-driven proton flux to increase as more of the excited electrons from PSI are used in cyclic electron flow to pump protons into the thylakoid lumen. The researchers found that the proton flux was 2–3 times higher in *hcef* plants.

      (c)   PSII activity was not affected because the altered electron flow happened after electrons had left PSII.

      (d)   An increased level of a compound often indicates impaired function of the enzyme using it in a pathway, in this case for glucose synthesis (see Figure 9–11). Fructose-1,6-bisphosphatase was found to be defective, resulting in an accumulation of its substrate fructose-1,6-phosphate.

      (e)   Starch synthesis was not measured in this experiment, but you would expect it to decrease due to the reduced synthesis of glucose resulting from defective fructose-1,6-bisphosphatase.

11-3. Here are three possible reasons:

      (1)   ATP is highly polar and would be difficult to move across membranes.

      (2)   ATP would be a very inefficient way to move energy about, because it has a molecular weight of about 500 daltons and contains only two high-energy phosphate bonds. Sucrose, in contrast, has a molecular weight of about 340 daltons and can give rise to 72 molecules of ATP elsewhere in the plant.

      (3)   Generation of ATP stops as soon as the sun sets, and there is no practical way to store enough of it to meet the ongoing energy needs of the organism at night.

11-4.   (a)   PGA down, RuBP up. With little or no carbon dioxide on hand to generate PGA from RuBP, the former will decrease in concentration and the latter will accumulate.

(b)   PGA up, RuBP down. In the presence of green light only, the light-dependent reactions will function minimally if at all, which means that there will be little or no ATP and NADPH to drive the reduction of PGA to glyceraldehyde-3-phosphate from which RuBP is regenerated. As a result, PGA will accumulate and RuBP will be depleted.

(c)   PGA up, RuBP down. The same reasoning as in part b; the products of the light-dependent reactions are required to keep the Calvin cycle functioning, and an inhibitor of photosystem II will effectively prevent the formation of NADPH and will almost certainly reduce the formation of ATP as well.

(d)   PGA up, RuBP unchanged. Decreased oxygen availability will lessen the oxygenase activity of rubisco, thereby allowing more net carboxylation and hence more PGA formation. RuBP consumption, on the other hand, is not likely to change; it is simply being used more for carboxylation and less for oxygenation.

11-5.   The flow of energy is as follows:

photon → excited electron in accessory pigment → excited electron in chlorophyll molecule → excited electron in special chlorophyll molecule at the reaction center of a photosystem → reduced organic electron acceptor → reduced intermediates along the electron transport system → electrochemical proton gradient across the thylakoid membrane → ATP → high-energy phosphate bond of glycerate-1,3-bisphosphate → glyceraldehyde-3-phosphate → fructose-6-phosphate → intermediate monosaccharides → starch.

11-6.   Although both the mint sprig and the mouse require air, they do not depend on it for the same reason. The mint sprig needs carbon dioxide, and the mouse needs oxygen. In fact, the photosynthetic activity of the mint sprig enhances the supply of oxygen in the air, rendering it "not at all inconvenient" to the mouse. The observation was important in establishing that whatever the photosynthetic organism acquired from air, it was obviously not the same material that a respiring animal depended on.

11-7.   Here are four possible reasons for the discrepancy:

(1)   Not all light will be of an appropriate wavelength (i.e., green light is reflected).

(2)   Many leaves may be shaded by leaves in the canopy above them.

(3)   The difference in energy between photons absorbed by light-harvesting pigments and the energy that actually excites an electron within the reaction center of a photosystem is lost as heat and entropy.

(4)   The capacity of a photosystem to absorb energy may be saturated by a level of light far below the intensity of sunlight.

11-8. (a)   Stroma.

(b)   Through photosystem I and the cytochrome b_6/f complex.

(c)   Stroma.

(d)   Nonappressed regions of the thylakoid membrane.

(e)   Lipid phase of the thylakoid membrane.

(f)   Across the thylakoid membrane from the stroma into the lumen.

(g)   Photosystem I.

(h)   Stroma.

(i)   Photosystems and light-harvesting complexes.

(j)   Photosystem II.

11-9. (a)   Across the outer and inner chloroplast membranes from the cytosol to the stroma.

(b)   Across the outer and inner chloroplast membranes from the cytosol to the stroma.

(c)   No.

(d)   No.

(e)   Across the inner and outer chloroplast membranes from the stroma to the cytosol.

(f)   No.

(g)   No.

(h)   Across the thylakoid membrane from the lumen to the stroma, and across the inner and outer chloroplast membranes from the stroma to the cytosol.

(i)   Across the thylakoid membrane from the stroma to the lumen.

(j)   *In the bundle sheath cells of C_4 plants:* across the inner and outer chloroplast membranes from the stroma to the cytosol.
In the mesophyll cells of C_4 plants: across the outer and inner chloroplast membranes from the cytosol to the stroma.

11-10. *During the night:* CO_2 in atmosphere → stomata → mesophyll cell → bicarbonate → oxaloacetate (in cytosol) → malate (in cytosol) → malate (in vacuole).
During the day: malate (in vacuole) → malate (in cytosol) → carbon dioxide (in cytosol) → Calvin cycle → glyceraldehyde-3-phosphate.
CAM plants are able to fix CO_2 and accumulate carbon in the form of organic acids at night, when the temperature is likely to be cooler and the loss of water through open

stomata thereby minimized. Open stomata are necessary for gas exchange between the interior and exterior of the leaf. During the day, when solar energy is available but the potential for water loss is highest, CAM plants close their stomata and channel carbon from malate into the Calvin cycle. C_3 and C_4 plants, on the other hand, must open their stomata during the day to acquire sufficient carbon dioxide.

11-11. (a)   The inner membrane systems found in mitochondria and chloroplasts may be derived from ingested bacteria, and the outer membrane systems may be derived from nucleated hosts. An identical arrangement of membranes follows phagocytosis: the ingested bacterium is surrounded by the membrane of the phagocytic vacuole. The thylakoid membrane systems found in chloroplasts may also be derived from ingested bacteria, because thylakoids form from invaginations of the inner membrane that pinch off to generate a separate membrane system.

(b)   One of the most important features the bacteria might have dispensed with is the cell wall. Without cell walls, the bacteria might burst due to osmotic pressure or might be more vulnerable to invasion by viruses.

(c)   See part b.

(d)   Features that mitochondria have retained while peroxisomes have not include (i) DNA replication, (ii) RNA synthesis, (iii) protein synthesis, (iv) an inner as well as an outer membrane, (v) electron transport, and (vi) ATP synthesis. Peroxisomes might have enabled ancient cells to (i) expand the use of oxidases—which generate the potentially toxic compound H_2O_2—for a variety of metabolic processes; (ii) generate acetyl coenzyme A from long-chain fatty acids, providing additional carbon sources for the cell; and (iii) reduce the toxicity of a variety of compounds by using the molecules as electron donors for breaking down H_2O_2.

12

The Endomembrane System and Peroxisomes

12-1. (a) Both peroxisomes and mitochondria. The process of β oxidation of long-chain fatty acids begins in the peroxisome but once the fatty acids have been shortened sufficiently in chain length, their further oxidation occurs in the mitochondrion. In the peroxisome, β oxidation generates NADH and acetyl CoA, which are essential for other peroxisomal reactions. In the mitochondrion, the acetyl CoA is immediately available for further oxidation by the TCA cycle.

(b) ER. Cholesterol is synthesized in the ER, where it is immediately available for steroid hormone biosynthesis or is easily transported throughout the endomembrane system.

(c) Rough ER and Golgi. Once in the ER, insulin can enter the secretory pathway for export from the cell.

(d) Smooth ER. Once in the ER, testosterone can enter the secretory pathway for export from the cell.

(e) Lysosome. Because the lytic enzymes that are needed for autophagy are sequestered in a membrane-bounded organelle, other cellular components are protected from degradation.

(f) Rough ER and Golgi. Membrane proteins can be glycosylated by the same general pathway used for glycosylating proteins destined for different locations within the cell, with variations in glycosylation used as targeting signals.

(g) Smooth ER. Phenobarbital is hydrophobic and therefore accumulates in the lipid portion of the smooth ER membrane; hydroxylation increases the solubility of the molecule in water, allowing the cell to export it via a secretory pathway.

(h) Golgi. The proteins are sorted after undergoing glycosylation, folding, and additional processing steps, thereby enabling the cell to use the same general pathway for processing proteins that are then targeted to different organelles.

12-2. (a) R, S (d) S (g) S

(b) R (e) R (h) R

(c) S (f) R (i) R, S

12-3. (a) For the pathway whereby glycoproteins of the plasma membrane are synthesized and glycosylated, see Figure S12-1.

(b) Glycoproteins are always found on the outer phospholipid monolayer of the plasma membrane with their carbohydrate side chains exposed on the outer surface because this is the monolayer that originally faced the interior of the rough ER and Golgi, where the enzymes involved in glycosylation are located.

(c) The answer to parts a and b require the assumptions that (i) membrane asymmetry is maintained throughout the rough ER, Golgi, and plasma membrane; and (ii) the relationship of these membrane systems to each other is essentially as illustrated in Figure S12-1.

12-4. (a) C (d) C, I, II (g) I (j) C, I, II

(b) II (e) C (h) C (k) I, II

(c) I (f) N (i) C

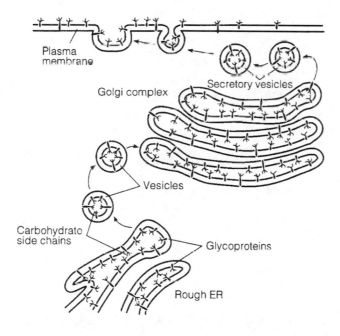

Figure S12-1 Synthesis and Glycosylation of Integral Membrane Proteins of the Plasma Membrane. Integral membrane proteins are synthesized on the rough ER, with oligosaccharide side chains added in part on the lumenal side of the rough ER (core glycosylation) and in part on the lumenal side of the Golgi complex (terminal glycosylation). Side chains therefore face the interior of both organelles as well as the interior of transport vesicles and become oriented toward the exterior of the cell when the vesicles fuse with the plasma membrane. See Problem 12-3.

12-5. (a) Colcichine is a drug that prevents microtubule assembly, thereby disrupting microtubule-based functions (see Chapter 15). This observation indicates the involvement of microtubules in intracellular movement of exocytic and endocytic vesicles.

(b) This finding suggests that the pathways for constitutive and regulated secretion, though separate, can occur simultaneously in the same cell.

(c) This result suggests that exocytic secretion may be triggered in vivo by an elevation of the intracellular Ca^{2+} concentration.

(d) This observation indicates that dynamin is essential for receptor-mediated endocytosis. However, there are also dynamin-independent pathways for ingesting extracellular fluid. When receptor-mediated endocytosis is inhibited, the rate of ingestion of extracellular fluid by other endocytic pathways increases.

(e) Because brefeldin A disrupts COPI-mediated vesicle transport, the results show that cholesterol efflux by exocytosis requires COPI-containing vesicles. Also, the results show that endocytosis and resecretion of apolipoprotein A-1 does not require COPI, and that altered trafficking of apolipoprotein A-1 cannot be the reason for the block in cholesterol efflux.

12-6. (a) P, R, A, E (d) A (g) P, R, A, E (j) P, R, A

(b) A (e) E (h) P, R (k) P, R, A

(c) P, R, E (f) P, A, E (i) E

12-7. (a) N; peroxisomes acquire proteins from ribosomes found in the cytosol.

(b) S; true only of glyoxysomes and some animal peroxisomes.

(c) N; acid hydrolases are found in lysosomes, not peroxisomes.

(d) A; catalase is a component of all peroxisomes.

(e) A; general property of peroxisomes.

(f) N; fireflies and related bioluminescent organisms have luciferase protein present in their peroxisomes but DNA is not found in this organelle.

(g) S; urate oxidase is present in many animal peroxisomes and in some, but not all, plant peroxisomes.

(h) A; peroxisomes are a source of as much as 50% of the dolichol found in cells.

(i) A; general property of peroxisomes.

12-8. (a) M (c) S (e) S (g) M

(b) S (d) N (f) S (h) N

12-9.   (1)   For the short peptide Lys-Asp-Glu-Leu:

(a)   Incorporated into ER-specific proteins as the polypeptides are synthesized by ribosomes attached to the rough ER.

(b)   When such proteins reach the Golgi complex, they bind to specific receptors and are packaged into transport vesicles for return to the ER.

(c)   Removal of this tag from ER-specific proteins usually results in their secretion from the cell.

(2)   For hydrophobic membrane-spanning domains:

(a)   Incorporated into membrane-bound Golgi complex proteins as the polypeptides are synthesized by ribosomes attached to the rough ER.

(b)   Such proteins tend to move through the endomembrane system until the thickness of the membrane, which increases progressively from the ER to the plasma membrane, exceeds the length of their membrane-spanning domains and blocks further migration.

(c)   Removal of the membrane-spanning domains would convert the membrane-bound proteins to soluble proteins and would also severely affect folding of the polypeptides; such abnormal proteins will probably be exported from the ER and degraded.

(3)   For mannose-6-phosphate residues:

(a)   Mannose-6-phosphate (M-6-P) residues are attached to soluble lysosomal proteins by the sequential action of several enzymes found in the lumen of the rough ER and Golgi complex.

(b)   When such proteins reach the TGN, they bind to M-6-P receptors and are packaged into vesicles for transport to endosomes.

(c)   Removal of M-6-P residues from lysosomal proteins often leads to secretion of the proteins; however, evidence suggests that additional tags may ensure delivery of some lysosomal proteins to endosomes.

12-10. (a)   The fibers or particles are probably taken up by endocytosis, followed by transport via early and late endosomes to a heterophagic lysosome.

(b)   The fibers or particles may physically abrade the lysosomal membrane, causing it to become leaky.

(c)   Cell death is probably due to the digestion of cellular components by acid hydrolases that escape from damaged lysosomes.

(d)   The fibers or particles released on cell death presumably are available for ingestion by other macrophages, with the same end result. Because the fibers and particles are not digestible and there is no mechanism to remove them from the lungs, a cycle of uptake, lysosomal damage, cell death, and fiber or particle release is set up that can continue indefinitely, killing more and more cells.

(e)   Exposure of silica particles apparently causes the macrophages to release a soluble factor that stimulates fibroblast cells in the lung to deposit collagen fibers, probably in an attempt to seal off and thereby contain the silica in the lung.

12-11. (a)   This observation is consistent with the prediction that the different plant exocyst proteins are associated in a multiprotein complex similar to the exocyst complex of yeast and mammals.

(b)   A mutation resulting in a defective or missing protein interferes with production of the functional multiprotein complex that is required for vesicle transport and subsequent germination.

(c)   This observation suggests that the different plant proteins work together and that all are required for proper function.

(d)   Because the growing tip is the site of expansion of the cell, this result suggests that the plant proteins are helping to supply material that is being added to the surface of the cell via exocytosis and membrane fusion.

(e)   Exocytosis of material to expand the growing tip has been blocked, suggesting that the exocyst proteins are required for incorporation of material into the cell surface by vesicle fusion.

13

Signal Transduction Mechanisms:
I. Electrical Signals in Nerve Cells

13-1. (a)   S   (c)   S   (e)   S   (g)   N

       (b)   N   (d)   N   (f)   S

13-2. (a)   These are the only three ions to which the plasma membrane of the nerve cell is sufficiently permeable to warrant their inclusion in the equation.

       (b)   A more general expression for monovalent ions is

$$V_m = \frac{RT}{F} \ln \frac{\sum P_{cation}[cation]_{outside} + \sum P_{anion}[anion]_{inside}}{\sum P_{cation}[cation]_{inside} + \sum P_{anion}[anion]_{outside}}$$

       (c)   With the relative permeability for sodium ions at 0.01, the value for V_m is –77 mV, calculated according to Equation 13-3 in the textbook. If the relative permeability for sodium ions were 1.0 instead, the value for V_m would be about –6.9 mV.

       (d)   No, because V_m is proportional to the logarithm of the expression that contains the term for sodium permeability.

13-3. (a)   (5 pA) (1×10^{-12} ampere/pA) (6.2×10^{18} charges/amp-sec) (0.005 sec) = **155,000 charges.** Thus, 155,000 ions pass through a single channel during the 5 milliseconds that it is open.

       (b)   No. Even though 155,000 ions pass through a single open channel, this is insufficient current to cause any perceptible change in the resting potential of a postsynaptic membrane. (An actual synaptic transmission event involves the simultaneous fusion of several hundred synaptic vesicles with the presynaptic membrane. Each vesicle releases several thousand neurotransmitter molecules, and each of these, in turn, causes thousands of receptor channels in the post-synaptic membrane to open for a few milliseconds. Thus, several hundred thousand channels open at once, generating a current that is sufficient to drive the resting potential of the postsynaptic membrane to its threshold, leading to depolarization of the membrane and an ensuing action potential.)

13-4. (a) E_{Cl} will be negative, because only a negative charge on the inner surface of the membrane would counteract the tendency of the concentration gradient to drive chloride ions inward.

(b) $E_{Cl} = (RT/zF) \ln \dfrac{[Cl^-]_{outside}}{[Cl^-]_{inside}} = -58 \log_{10} \dfrac{560}{50} = \mathbf{-60\ mV.}$

(Note that the RT/zF value is negative because $z = -1$ for an anion.)

(c)   If the internal chloride ion concentration is 150 mM instead of 50 mM, the value of E_{Cl} becomes less negative (-33 mV instead of -60 mV).

13-5.   (a) $E_K = \dfrac{(2.303)(1.987)(273+37)}{(+1)(23,062)} \log_{10} \dfrac{4.6}{150}$

         $= +0.06157 \log_{10}(0.0306)$

         $= -0.0932\ V = \mathbf{-93.2\ mV.}$

   $E_{Na} = +0.06157 \log_{10} \dfrac{145}{10} = +0.0715\ V = \mathbf{+71.5\ mV.}$

   $E_{Ca} = (+0.06157/2) \log_{10} \dfrac{6}{0.001}$

         $= +0.0308 \log_{10}(6000)$

         $= +0.116\ V = \mathbf{+116\ mV.}$

(b)   The resting potential of a cardiac muscle cell is substantially more negative than the squid axon because of a greater potassium ion concentration gradient across the membrane of the cardiac cell.

(c)   Either Na^+ or Ca^{2+}, or both; both would move *inward*.

(d)   You could remove either sodium or calcium ions from the surrounding medium and observe the effect this has on the action potential. Radioactive isotopes could also be used to follow specific ions.

(e)   Potassium is driven outward both by its concentration gradient across the membrane and by the temporarily positive membrane potential. At A, the voltage-dependent potassium gates are not yet open.

13-6.   (a)   Of the sodium channels in the membrane, at least some will open in response to a stimulus that depolarizes the membrane by about 20 mV.

(b)   Intensity of stimulus is detected as the frequency with which individual neurons respond and/or as the difference in the number of separate neurons that respond.

13-7.   An action potential cannot move backwards in the direction of a membrane that has just experienced an action potential. Once a region of the membrane has experienced an action potential, it becomes temporarily unresponsive and incapable of undergoing another action potential. This is known as the refractory period. The refractory period can be divided into two periods known as the absolute and relative refractory periods. The absolute refractory period is due to the inactivation of sodium channels. Following the absolute refractory period is the relative refractory period where a new action potential is possible but difficult to initiate. The relative refractory period is due to the increased permeability to potassium ions which follows the opening of sodium channels during an action potential.

13-8. Saltatory propagation would be disrupted or slowed at the very least, because the insulating layer of myelin would be reduced or absent. In addition, voltage-gated channels are clustered at nodes of Ranvier in myelinated neurons. Loss of saltatory propagation might result in the inability to propagate action potential as far as the next node in the absence of insulating myelin.

13-9. Increasing the concentration of potassium ions should hyperpolarize neuronal membranes, causing the resting membrane potential to be more negative and thus making it more difficult to depolarize the membrane to threshold. This should reduce the frequency of muscle stimulation.

13-10. (a) Reduction of calcium influx at the presynaptic nerve terminal would result in decreased exocytosis of vesicles containing glutamate from the nerve terminal.

(b) Hyperpolarization has an inhibitory effect on action potential generation in the postsynaptic neuron, because it makes it more difficult for the postsynaptic nerve terminal's membrane to reach the threshold for action potential firing. Since THC reduces glutamate release, the postsynaptic neuron will be *more* likely to fire, and so it will be overstimulated.

13-11. If an excitatory neurotransmitter remains in the synaptic cleft for prolonged periods of time, the postsynaptic neuron will experience an excitatory postsynaptic potential (EPSP) for a longer period of time than normal. This will result in a prolonged period during which a rapid succession of action potential will be generated in the postsynaptic neuron.

CHAPTER

14

Signal Transduction Mechanisms: II. Messengers and Receptors

14-1.   (a)   Calmodulin

      (b)   endocrine

      (c)   ligand

      (d)   inositol trisphosphate (IP_3) and diacylglycerol (DAG)

      (e)   adenylyl cyclase; phosphodiesterase

      (f)   endoplasmic reticulum

14-2.   (a)   The activation of Ras requires the assistance of another protein called SOS. Upon binding to the tyrosine kinase receptor, SOS catalyzes the Ras GDP/GTP exchange reaction. In this respect, SOS and Ras together are similar to a heterotrimeric G protein. Here, SOS acts like the $G\beta\gamma$ subunits, whereas Ras is similar to a $G\alpha$ subunit.

      (b)   Activation of heterotrimeric G proteins involves a conformational change in a protein with which it associates (a receptor). Ras activation is stimulated by a conformational change in a guanine nucleotide exchange factor (GEF), such as SOS. Inactivation of the two types of G proteins is similar as well. Regulator of G protein signaling (RGS) proteins increase the rate of hydrolysis of GTP by heterotrimeric G proteins, leading to their inactivation. These are similar to GTPase activating proteins (GAPs), which increase the rate of Ras inactivation.

14-3.   The chelator could be added to the extracellular medium to reduce the calcium outside the cell. Addition of the chelator to the extracellular medium should block the action of the hormone if it depends on calcium influx. If the action of the hormone depends solely on calcium release from intracellular stores such as the endoplasmic reticulum, then addition of a calcium ionophore should mimic the action of the hormone. Furthermore, the ionophore should be effective even when EGTA is added to the extracellular medium to reduce the free calcium ion concentration.

14-4.   (a)   These receptors cannot recruit Grb2 and SOS, so they will not transduce a signal, and will not differentiate as R7 cells properly.

      (b)   Ras will now trigger differentiation even if the R7 cannot transduce the normal signal due to lack of functional SOS. The result will be that R7 cells do form. (In fact, in this situation an excess of R7 cells sometimes results.)

(c) Without functional MAPK, signals will not be transduced to the nucleus, so a lack of R7 cells would be expected.

(d) Even if the R8 cells produce too much signal, R7 cells lacking the receptor will be "blind" to the excess signal, resulting in loss of R7 cells.

14-5. The binding of epinephrine to β-adrenergic receptors causes a stimulation of heart function, both in terms of heart rate and with respect to the amount of work done in pumping blood. The effect appears to be mediated by cyclic AMP. When an antagonist such as the beta blocker propranolol is given to patients with hypertension, the cellular response caused by the binding of epinephrine to beta receptors is partially inhibited. Heart function is gradually restored over a period of time, with a corresponding decrease in blood pressure.

14-6. (a) It takes approximately 1.5–2 min to reach maximal Ras activation.

(b) Ras activity gradually decreases due to the action of Ras GAPs, which stimulate Ras to hydrolyze GTP to GDP, inactivating Ras.

14-7. Pertussis toxin ADP ribosylates certain G proteins. It would be likely that a neutrophil responds to bacterial proteins through a G protein-linked receptor.

14-8. The response to smoking is complex and involves a large number of changes in the function of neurons. However, smokers do develop tolerance to nicotine. Therefore, it would be reasonable to suggest that chronic exposure to nicotine might lead to receptor down-regulation as a basis for this tolerance.

14-9. (a) Uncaging calcium ions locally should act like the normal increase in calcium that accompanies fertilization (assuming the starfish oocyte has been matured in vitro). This should initiate a calcium wave.

(b) If these receptors were made in sufficient quantities, they would "compete" with normal receptors for available IP_3 (i.e., they would act in dominant-negative fashion). The result would be reduced calcium release from the ER following fertilization.

(c) A wave of aequorin fluorescence would be observed to sweep across the egg, initiating at the site of sperm entry into the egg.

15

Cytoskeletal Systems

15-1. (a)   MF   (c)   MT   (e)   IF   (g)   MF, MT   (i)   MF, MT
      (b)   MT   (d)   MF   (f)   MF, MT, IF   (h)   MF, MT, IF   (j)   MT

15-2. (a)   F. The *rate* of loss is greater than the *rate* of addition at the minus end over time. There is no absolute absence of addition at the minus end.

      (b)   F; hydrolysis of the ATP bound to actin and the GTP bound to tubulin usually occurs during monomer polymerization, but the polymerization process still occurs even if the ATP or GTP is replaced by a nonhydrolyzable analogue.

      (c)   T; although MTs and MFs exchange subunits much more dynamically than IFs, there is still some assembly and disassembly that occurs with IFs.

      (d)   T; *Listeria* "rocketing" requires extensive polymerization of actin behind the bacterium as it moves within infected cells.

      (e)   F; algae are eukaryotes and therefore possess the same kinds of cytoskeletal components as other eukaryotic cells.

      (f)   T

      (g)   F. Most do, but not all. In addition, in ciliated or flagellated cells, the minus ends of microtubules can be anchored at the basal body, which also serves as a microtubule-organizing center.

      (h)   M; the statement is true if the monomer concentration is above the overall critical concentration but false otherwise.

15-3. (a)   Pigment granule dispersal is a microtubule-dependent process.

      (b)   The centrosome serves as a microtubule-organizing center in vivo, and all of the microtubules radiating from the centrosome apparently have the same polarity.

      (c)   The extracts appear to contain structures that are functionally equivalent to centrosomes (as evidenced by the presence of γ-tubulin), which nucleate microtubule growth.

15-4.   The overall critical concentration should be reduced when tubulin polymerizes in the presence of the centrosome preparation. If the minus end cannot disassemble, the overall critical concentration should now be equivalent to the lower critical concentration of the plus end.

15-5.   (a)   3. Without the aid of Arp2/3 or WASP family proteins, polymerization is very slow.

   (b)   2. Arp2/3 aids polymerization, but not as much as Arp2/3 + WASP family proteins.

   (c)   1. Arp2/3 + WASP family proteins cause explosive polymerization of actin at essentially maximal rates.

15-6.   Cytochalasin D prevents addition of new monomers to existing filaments, effectively capping them. In situations in which the concentration of G-actin in the cytosol is below the critical concentration for net addition of subunits to the minus end of the capped filaments, loss of monomers at the minus end of existing filaments eventually results in their shortening. This occurs despite the pool of available G-actin in the cytosol.

15-7.   (a)   Gelsolin severs microfilaments, resulting in loss of tension-generating elements, so wrinkling should decrease.

   (b)   Decreasing Rho activity by adding C3 transferase should decrease contractility by decreasing the number of stress fibers. This causes a reduction in wrinkles.

   (c)   Taxol stabilizes microtubules, which are thought to bear compressive loads. Now the cells should be better able to resist their own tension, and the wrinkling of the rubber should be less pronounced.

   (d)   Latrunculin A would cause the loss of F-actin, resulting in the same effects as in (a). After the drug is washed out, contactility should eventually resume, as sufficient microfilaments form to allow force production.

   (e)   A constitutively active Rho would cause excess contraction, leading to more wrinkles.

15-8.   Disrupting intermediate filaments will result in cells that are more susceptible to mechanical forces. In the case of the keratinocytes, disrupting keratin, a key IF in these cells, would result in very fragile cells. Less force would need to be applied using the magnetic beads to damage them or change their shape.

16

Cellular Movement: Motility and Contractility

16-1. (a) The nonmotility of the sperm cells (and hence the sterility of the individual) is due to a structural defect in the outer dynein arms of the sperm tail axoneme, causing the sperm tail (or flagellum) to be nonfunctional. (In addition to the defect underlying Kartagener's triad, there are several other defects that also result in sperm nonmotility and hence male infertility. These include a defect in, or lack of, any of the following: both dynein arms, the inner arm only, the radial spoke heads, or one or both microtubules of the central pair.)

(b) The same structural defects in the sperm tail are also likely to affect the cilia responsible for sweeping mucus and foreign matter out of the lungs and sinuses.

16-2. (a) A, H   (c) A, H, I, R   (e) A, H, I   (g) A
(b) R   (d) I   (f) H

16-3. (a) At a sarcomere length of 3.2 µm, the length of the A band is 1.6 µm, and the length of the I band is also 1.6 µm, with 0.8 µm on either side of the Z line. During contraction of the sarcomere to 2.0 µm, the length of the A band remains fixed at 1.6 µm, and the length of the I band decreases from 1.6 µm to 0.4 µm.

(b) The H zone corresponds to that portion of the thick filament length that is not overlapped by thin filaments. (It is, in fact, the lack of interdigitated thin filaments that gives the H zone its lighter density and hence its German name.)

(c) The distance from the Z line to the edge of the H zone represents the length of the thin filament and remains constant during contraction.

16-4. (a) Rigor results from a failure to break the cross-bridges that link thick filaments to thin filaments in the contraction cycle. In the living cell, detachment occurs upon binding of the next molecule of ATP. After death, however, the supply of cellular ATP is quickly depleted and cannot be restored. This has two consequences: (1) ATP is not available to cause cross-bridge detachment; and (2) in the absence of ATP, calcium cannot be pumped into the sarcoplasmic reticulum. Calcium therefore accumulates in the cytosol and promotes attachment of cross-bridges to actin. The net result is an accumulation of cross-bridges that lock the muscle filaments together and give the corpse its characteristic stiffness.

(b)   While running to class; the concentration of ATP in the cell would presumably be lower and cross-bridge formation would probably occur more quickly upon death than if you were just sitting in lecture.

(c)   Addition of ATP will have a relaxing effect, because it will allow cross-bridges to break and filaments to detach.

16-5.   (a)   The cross-bridges will be dissociated from the thin filaments but will not be "recocked" to the conformation normally seen in resting muscle, because AMPPCP cannot be hydrolyzed to ADP and P_i. Your diagram should look like that after step 3 in Figure 16-18 on p. 465 of the textbook, except that AMPPCP is bound to the myosin head instead of ATP.

(b)   Yes, because as a structural analogue, AMPPCP is presumably a competitive inhibitor of the ATPase and should be displaced from the binding site if the intracellular ATP concentration is elevated to a high enough level.

(c)   All other ATP-dependent processes are likely to be inhibited, including uptake of calcium ions by the sarcoplasmic reticulum.

16-6.   (a)   The myosin thick filaments are assembled with the rodlike tail domain in the center of the filament and the globular heads located at the ends of the filament pointing away from the center. Therefore, the globular myosin heads at each end of the thick filament have opposite polarities. This results in the movement of the actin filaments in opposite directions so that actin filaments from each Z line are drawn to the center of the sarcomere.

(b)   For stress fibers to exert tension, there would have to be regions of antiparallel actin filaments—that is, regions in which some actin filaments attached to myosin have opposite orientation. The net effect of such an arrangement is to create a disordered "mini-sarcomere," in which the movement of the actin filaments in opposite directions causes the stress fiber to shorten toward its center.

16-7.   Because reuptake of Ach is inhibited, the muscle membrane at the neuromuscular junction will remain depolarized for an abnormally long period, causing the muscles to remain in a contracted state. The prolonged depolarization, in turn, would be propagated via the T tubule system to the sarcoplasmic reticulum. The result would be prolonged release of calcium from the SR, resulting in sarcomere shortening via the effects of calcium on the troponin system.

16-8.   (a)   Particles use different motors to move outward and inward. Each motor has a characteristic velocity, which accounts for the difference.

(b)   IFT seems to be necessary for ongoing maintenance of flagellar structure, presumably by transporting new components to the tips of flagella to replace components that are turning over.

(c)   Kinesins transport cargoes toward the plus ends of microtubules, so the plus ends should be at the tip of the flagellum.

16-9. (a) AMPPNP would ultimately inhibit flagellar dynein, causing cessation of flagellar bending.

(b) The molecule isolated using this procedure was classical kinesin.

(c) In order to get the wave-like motion seen in the beating of sperm flagella, some dynein molecules must be exerting force while those on the opposite side of the tail must be inactive. If dynein molecules were simultaneously exerting force on all sides, the sperm tail would be locked in a rigid state, unable to move at all.

CHAPTER

17

Beyond the Cell:
Extracellular Structures,
Cell Adhesion, and Cell Junctions

17-1. (a) Both the ECM of animal cells and the walls around plant cells consist of long, rigid fibers embedded in an amorphous, hydrated matrix of branched-chain molecules, either glycoproteins (ECM) or polysaccharides (cell wall).

(b) For ECM, the fibers consist of collagen and the matrix is a network of proteoglycans. For cell walls, the fibers consist of cellulose and the matrix is a network of polysaccharides and proteins.

(c) Both ECM and cell walls are important in maintaining cell shape and in retaining water, thereby resisting compression.

(d) Roles unique to the ECM include regulation of cellular processes such as adhesion, motility, and differentiation during embryonic development. Roles unique to cell walls include protection of the cell from mechanical injury and microbial invasion, as well as provision of the mechanical support necessary to withstand the turgor pressure that gives plant tissues their rigidity. Some animal extracellular structures, such as the exoskeletons or cuticles of some invertebrates, also provide support and fight microbial invasion.

17-2. (a) Both fibronectin (FN) and laminin (LN) have multiple domains that allow them to attach to other ECM proteins. In the case of fibronectin, it can bind to fibrin (important during blood clot formation), and can bind to collagen or heparan sulfate proteoglycans for attaching it to the ECM. The RGD-containing cell-binding domain of FN is crucial for allowing cells to attach to FN, because this is the site that is bound by integrins on the cell surface.
In the case of LN, there are similar domains. A collagen IV binding site allows it to be attached to basal laminae. LN also has an integrin-binding site, although in the case of LN, there is a different set of integrin α- and β-subunits.

(b) RGD peptides, added in sufficient quantities, will tend to bind to the integrin receptors on the surface of cells. If integrins are already bound to these peptides, their receptors will be unavailable for binding to FN (or LN), and thus cells will be inhibited from binding to the ECM. This sort of "competition" experiment has often been used to show that a cellular process is integrin-dependent.

17-3. (a) Collagen/elastin: Collagen is responsible for the strength of the ECM, whereas elastin imparts elasticity and flexibility to the ECM.

(b) Fibronectin/laminin: These are the two most common kinds of adhesive glycoproteins; fibronectins occur widely throughout supporting tissues and body fluids, whereas laminins are found mainly in the basal laminae.

(c) Integrin/selectin: The integrins are transmembrane proteins that serve as receptors for ECM proteins such as collagen, fibronectin, and laminin; selectin is the best-characterized integrin.

(d) IgSF/cadherin: Both IgSFs and cadherins mediate cell–cell recognition and adhesion; the two groups of proteins can be distinguished from each other functionally because of the calcium requirement of cadherins but not IgSFs.

(e) ECM/glycocalyx: The glycocalyx is a carbohydrate-rich zone juxtaposed between the plasma membrane and the ECM of many types of animal cells.

(f) Hemidesmosome/focal adhesion: Both are adhesive junctions that connect cells to the ECM, with integrins as their main transmembrane linker proteins.

(g) Apical surface/basolateral surface: The surfaces of intestinal epithelial cells that face the lumen of the intestine and the circulatory system, respectively; maintained as separate domains with respect to integral membrane proteins because tight junctions prevent proteins on the apical surface from moving laterally in the membrane to the basolateral surface, and vice versa.

17-4. Compaction is probably mediated by cadherins, because the proteins involved appear to be calcium-sensitive cell surface proteins. In reality, the molecule involved is E-cadherin, which was also "uvomorulin" in the early literature.

17-5. (a) A   (c) G   (e) G, P

(b) T   (d) T

17-6. (a) G; forms channels (cytoplasmic connections) between adjacent animal cells.

(b) A; transmembrane adhesive connections between two adjacent animal cells.

(c) A; probably binds the plasma membranes of the two cells together.

(d) P; tubular structure found in the center of a plasmodesma.

(e) A; major proteins of the desmosome plaque.

(f) P; forms channels (cytoplasmic connections) between adjacent plant cells.

(g) A; anchors actin microfilaments to the plasma membrane.

(h) T; binds adjoining cells together, creating fused ridges.

17-7. (a) Gap junctions between cells will allow passage of molecules and ions up to a specific size limit. The gap junctions that connect adjacent cells in insect salivary glands allow the passage of small molecules (molecular weight up to at least 1158 daltons), whereas larger molecules (1926 daltons) do not flow between cells.

(b) The passage of molecules between cells can be regulated by the intracellular Ca^{2+} content. When the Ca^{2+} concentration is increased in an individual cell, the movement of fluorescent molecules into that cell is inhibited, suggesting that gap junctions are closed under these conditions.

17-8. Normally, the positively charged amino acids near the tip of the claudin cause electrostatic repulsion of Na^+ ions, which leads to low paracellular permeability through the tight junction. When these amino acids are changed to negatively charged amino acids, this repulsion is abolished, allowing the Na^+ ions to pass more easily through the tight junction via paracellular permeability.

17-9. (a) Primary cell walls are laid down as new cells are formed and are extensible under the appropriate conditions. In some plant cells, secondary cell walls are deposited on the inner surface of the primary wall. Because of their high cellulose and lignin contents, secondary cell walls are inextensible and therefore specify the final size and shape of the cell definitively.

(b) Cellulose composes the rigid microfibrils of the cell wall, whereas hemicellulose is one of the major polysaccharides that make up the amorphous matrix in which the cellulose microfibrils are embedded.

(c) Xylan is a polymer of xylose, whereas xyloglucan is a polymer of glucose with side chains that consist of xylose, galactose, arabinose, and fucose. Xylan is found in monocots and xyloglucan in dicots.

(d) Extensins are rigid, rodlike glycoproteins with numerous oligosaccharide side chains that are tightly integrated into the polysaccharide matrix of the cell wall. Lignins are polymers of aromatic alcohols that form cross-linked networks with extensin and other cell wall components.

(e) The desmotubule is the cylindrical structure in the central channel of a plasmodesma. The annulus is the ring of cytoplasm between the desmotubule and the membrane that lines the desmotubule.

(f) The plasmodesma is the junction that connects the cytoplasms of two adjacent plant cells. It plays the same role as the gap junction between animal cells.

17-10. (a) Hydroxyproline residues, but apparently not unhydroxylated proline residues, stabilize the collagen triple helix. (The hydroxyl groups of this amino acid form interchain hydrogen bonds that help stabilize the assembled triple-stranded helix.) If vitamin C (ascorbic acid) is deficient in the diet, tissue levels of ascorbic acid will be low, and the enzyme prolyl hydroxylase is not maintained in the reduced form and is therefore inactive. Proline is not hydroxylated, resulting in the inadequate stabilization of the triple helix. Collagen therefore breaks down, leading to defects in tissues that depend on collagen to maintain the adhesive strength of the ECM. As a consequence, such tissues are subject to breakdown and bruising. (Blood vessels also become fragile, which accounts for the hemorrhaging that is characteristic of scurvy.)

(b)   For the disease condition to become progressively worse—and for a patient with scurvy to respond to dietary vitamin C—collagen degradation and replacement must be relatively rapid, at least in susceptible tissues. (The turnover rate of collagen varies greatly, in fact; it is quite rapid in connective tissue but very slow in bone.)

(c)   Sailors on long sea voyages were susceptible to scurvy because no fresh fruit—and hence no source of vitamin C—was available. This is no longer a problem, because once the connection between vitamin C deficiency and scurvy was established, sailors were provided with citrus fruit. The term "limey" reflects the British Navy's use of limes for this purpose.

18

The Structural Basis of Cellular Information: DNA, Chromosomes, and the Nucleus

18-1.   (a)   It made their experiment possible, because it allowed them to assay various fractions of S cells for transforming activity in culture.

      (b)   This was the very hypothesis their experiment was designed to test; all that was left to do was to devise a means of identifying the nucleic acid if it did indeed "flow into the cell."

      (c)   The concept of base pairing by hydrogen bonding between pyrimidines and purines on opposite strands turned out to be a vital clue to the double-stranded structure of the DNA molecule.

      (d)   By showing that a virus was indeed like a "little hypodermic needle" capable of injecting its nucleic acid into the cell, Hershey and Chase were able to explain Anderson and Herriott's observation in terms of an osmotic shock that causes the viruses to empty their nucleic acid contents into the medium.

18-2.   (a)   Because A does not equal T, and G does not equal C, this cannot be a double-stranded DNA molecule. It must therefore be a single-stranded DNA.

      (b)   If 40% of the bases are G or C in a double-stranded DNA molecule, the remaining 60% must be A and T. In a DNA double helix, the amounts of A and T are equal to each other, so half of this 60%, or 30%, of the bases must be A.

      (c)   If 40% of the bases are G or T in a double-stranded DNA molecule, the remaining 60% must be A and C. But A and C do not form complementary base pairs with each other, so you don't know how much of this 60% is A and how much is C.

      (d)   If 15% of the bases are A in a double-stranded DNA molecule, then 15% must be the complementary base T, giving a combined total of 30% for A + T. The remaining 70% must be G and C. In a DNA double helix, the amounts of G and C are equal to each other, so half of this 70%, or 35%, must be the base C. According to the graph shown in Figure 18-9 on p. 517 of the textbook, a double-stranded DNA molecule containing 70% G + C would be expected to exhibit a T_m above 90°C.

18-3.   (a)   T   (c)   F   (e)   F

      (b)   T   (d)   T

18-4. Figure S18-1 depicts the phage DNA, which has a total of 10.5 kilobase pairs. The vertical arrows indicate sites on the DNA that are cut by restriction enzymes X and Y.

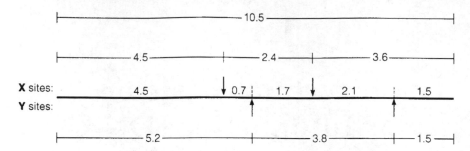

Figure S18-1 **Restriction Mapping.** Shown here is the restriction map of the bacteriophage DNA, with vertical arrows to indicate sites on the DNA that are cleaved by restriction enzymes X and Y. See Problem 18-4.

18-5. The DNA fragment to be analyzed is illustrated in Figure S18-2a, with the four-base primer region shown on the left and the unknown sequence on the right. The base sequence of the DNA fragment is determined by mixing together (1) a single-stranded (denatured) preparation of the DNA fragment; (2) the deoxynucleotides dATP, dCTP, dTTP, and dGTP; (3) the dideoxynucleotides ddATP, ddCTP, ddTTP, and ddGTP, each labeled with a fluorescent dye of a different color (e.g., ddATP = red, ddCTP = blue, ddTTP = orange, and ddGTP = green); (4) DNA polymerase; and (5) the single-stranded primer. After incubation, the reaction products are separated by gel electrophoresis. As indicated in Figure S18-2b, the pattern of colored bands in the gel will reveal the sequence of the original DNA.

(a)

```
3'-A-G-C-G-C-T-A-T-A-G-C-G-C-T-5'
5'-T-C-G-C-G-A-T-A-T-C-G-C-G-A-3'
```

Primer   Unknown sequence

(b)

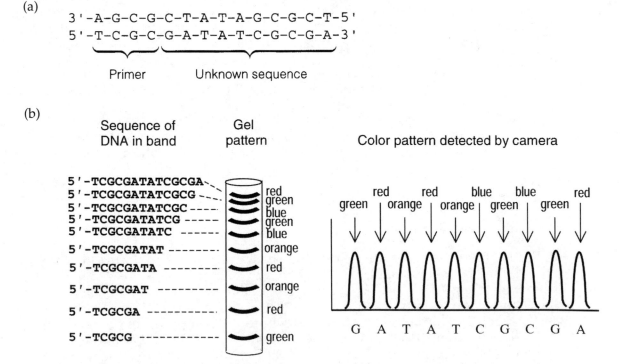

Figure S18-2 **DNA Sequencing.** (a) The DNA fragment to be analyzed. (b) The pattern of bands obtained using dye-labeled dideoxynucleotides, showing the base sequence of the DNA in each band and the pattern of colors detected by the camera. See Problem 18-5.

18-6.   (a)   Sample A has a higher T_m value, so it must have a higher content of G and C and a lower content of A and T than sample B.

         (b)   Sample A has less heterogeneity in base composition than sample B, possibly because genome A is smaller than genome B.

         (c)   Formamide and urea destabilize the DNA duplex by forming hydrogen bonds to the bases of either strand, causing the duplex structure to melt at a considerably lower temperature.

18-7.   (a)   If the two DNA samples had nucleotide sequences that were similar to each other, but not the same, strands from one sample could hybridize to strands from the other sample. Because the resulting hybrids would not be exactly complementary, you would get a lower melting temperature because DNA molecules in which the two strands of the double helix are properly base-paired at each position melt at higher temperatures than DNA in which the two strands are not perfectly complementary.

         (b)   To test your hypothesis, you might carry out base-sequencing analysis of the two DNA samples. If your hypothesis was correct, the two DNA base sequences should be similar, but not the same.

         (c)   There are two possible interpretations of such a result. First, the two initial samples of DNA might have been identical, in which case mixing together their individual strands would have no effect on the melting temperature following reassociation. Alternatively, the two initial samples may have been totally unrelated to each other in nucleotide sequence. In this case, strands from one sample could not hybridize with strands from the other sample during the reassociation step. Thus, the strands from sample "A" would only hybridize with complementary strands from sample "A" during the reassociation step, and strands from sample "B" would only hybridize with complementary strands from sample "B." Hence, the original melting temperature of each sample (92°C) would be retained after reassociation. The fact that both samples initially had the same melting temperature does not mean that they were related in sequence, although it does indicate that the relative proportion of GC base pairs in both samples was similar.

18-8.   (a)   The fact that the DNA fragments are all multiples of a basic unit 260 bp in length suggests that proteins are clustered along the DNA in a regular pattern that repeats at intervals of roughly 260 bp. Such a regular distribution of protein clusters suggests the existence of nucleosomes, even though the distance between them appears to be longer than the more typical value of 200 bp described in the chapter.

         (b)   In this case, each nucleosome appears to be associated with 260 bp of DNA.

         (c)   You would expect to see a series of fractions containing differing numbers of particles. The smallest fraction would contain single particles, the next smallest fraction would contain clusters of two particles, the succeeding fraction would contain clusters of three particles, and so forth (see Figure 18-19 on p. 531 of the textbook). If the DNA were isolated from these fractions and analyzed by gel electrophoresis, the DNA from the fraction containing single particles would be expected to measure 260 bp in length, the DNA from the fraction containing clusters of two particles would be expected to measure 520 bp in length, the DNA from the fraction containing clusters of three particles would be expected to measure 780 bp in length, and so on.

(d)   This experimental observation indicates that the nucleosomal core particle contains 146 bp of DNA. Because a total of 260 bp of DNA is associated with the nucleosome, the linker must be 260 − 146 = 114 bp.

18-9.   (a)   Unlike most membranes, the nuclear envelope appears to be freely permeable to a polar organic molecule.

(b)   The aqueous channels in nuclear pore complexes have diameters of at least 5.5 nm, but not as great as 15 nm.

(c)   The stained pores contain complexes of RNA and protein, probably ribonucleo-protein particles caught in transit.

(d)   The NLS only triggers transport from cytoplasm to nucleus, not from nucleus to cytoplasm, suggesting that NLS receptor proteins (importins) function only in the cytoplasm.

(e)   The nuclear membranes and the endoplasmic reticulum are likely to have a common origin.

(f)   Ribosomal proteins must pass inward from the cytoplasm to the nucleus at a rate adequate to sustain ribosomal subunit assembly; ribosomal subunits must move outward from the nucleus to the cytoplasm at a rate commensurate with their rate of assembly.

(g)   Nucleoli are probably responsible for rRNA synthesis.

(h)   The integrity of the nucleus does not seem to depend entirely on the envelope.

18-10.   NTF2 is required for recycling of Ran-GDP back into the nucleus. If the temperature is raised to the restrictive temperature, so that NTF2 stops functioning, Ran-GDP would be unable to return to the nucleus. The eventual result would be that Ran-GDP levels would drop in the nucleus. Without a supply of Ran-GDP, there would be reduced Ran-GTP available to shuttle proteins out of the nucleus. Thus, eventually, export of proteins out of the nucleus would stop.

18-11.   (a)   F; nucleoli are structures made of DNA, RNA, and protein that are present in the eukaryotic nucleus.

(b)   T

(c)   F; the DNA of nucleoli carries the cell's rRNA genes, which are present in cluster of multiple copies.

(d)   F; a single nucleolus may contain a number of NORs.

(e)   T

(f)   T

19

The Cell Cycle, DNA Replication, and Mitosis

19-1.   (a)   S   (d)   G1   (g)   G1, S, G2   (j)   G1, S, G2, M
        (b)   M   (e)   M   (h)   M
        (c)   M   (f)   G1, S, G2 (i)   G1, S, G2, M

19-2.   (a)   Mitotic index = (30 + 20 + 20 + 10 + 20)/1000 = 0.1 = **10%.**

        (b)   3% of time in prophase, 2% in prometaphase, 2% in metaphase, 1% in anaphase, 2% in telophase, 40% in G1, 30% in S, and 20% in G2.

        (c)   Because G2 is 20% of the cell cycle and lasts 4 hours, the whole cycle must be 4/0.2 = **20 hours.**

        (d)   0.6 h in prophase; 0.4 h each in prometaphase, metaphase, and telophase; 0.2 h in anaphase; 8 h in G1; 6 h in S; and 4 h in G2.

        (e)   The first appearance of label in prophase nuclei would have to be observed.

        (f)   About 30%, because all cells in S phase will incorporate label immediately, and on the average about 30% of the cells in the culture should be in S phase at any one time.

19-3   (a)   Figure S19-1a illustrates the expected progeny molecules for two rounds of DNA replication according to the conservative model (on left) and the dispersive model (on right).

        (b)   Figure S19-1b depicts the distribution of DNA bands in cesium chloride gradients after two rounds of DNA replication according to the conservative model (on left) and the dispersive model (on right).

(a)

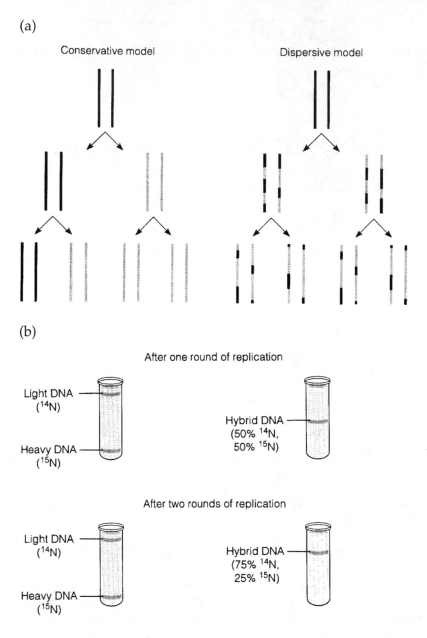

(b)

Figure S19-1   Meselson and Stahl Revisited. (a) The progeny molecules for two rounds of DNA replication according to the conservative (left) and dispersive (right) alternatives to semiconservative replication. (b) The distribution of DNA bands in cesium chloride gradients after one and two rounds of replication by the conservative (left) and dispersive (right) alternatives to semiconservative replication. See Problem 19-3. (It is precisely because Meselson and Stahl did not observe either of these patterns that these models were discarded in favor of the semiconservative model depicted in Figure 19-3 on p. 552 of the textbook.)

19-4.   Your sketch should look something like step 7 of Figure 19-13 on p. 563 of the textbook, with single-stranded DNA identified as the template, Okazaki fragments identified as newly forming pieces of DNA on the lagging strand, short segments of RNA on the Okazaki fragments identified as primers, and molecules of single-strand binding protein bound to the single-stranded DNA near the point at which the fork is opening. In addition, your sketch should include the following enzymes:

DNA helicase: Unwinds double-stranded DNA.

Gyrase: Nicks DNA ahead of the replication fork to relax the supercoiling induced in the DNA by helicase activity.

Primase: Synthesizes short segments of RNA used as primer for DNA synthesis.

DNA polymerase III: Elongates the growing segments of DNA.

DNA polymerase I: Removes RNA primer and replaces it with deoxyribonucleotides.

Ligase: Links discontinuous fragments of DNA covalently into a continuous strand.

19-5. (a) R; in *E. coli*, exonucleolytic activity is needed for RNA primer removal and for proofreading.

(b) R; in *E. coli*, synthesis is continuous on one strand and discontinuous (with Okazaki fragments) on the other.

(c) NB; the degree of sequence reiteration is not likely to have any effect on the mode of replication of the DNA.

(d) S; Okazaki fragments have short RNA (ribose) segments as well as longer DNA (deoxyribose) segments.

(e) R; this finding is not consistent with the semiconservative mode of replication found for *E. coli* DNA.

19-6. The cells have a defective gene for DNA ligase, causing the enzyme to lose its function at 42°C. DNA ligase is required during DNA replication for joining together the short Okazaki fragments that are created as part of lagging strand synthesis. The DNA denaturation step in your experiment released the two full-length strands of the original DNA, the new full-length leading strand, and the unjoined Okazaki fragments of the lagging strand. The two original full-length strands and the new leading strand together account for 75% of the DNA. These DNA strands have a molecular weight that is half that of the original DNA because they are single DNA strands rather than a DNA double helix. The remaining 25% of the DNA has a low molecular weight because it represents the unjoined Okazaki fragments of the lagging strand.

19-7. (a) Without origins of replication, the DNA would not be replicated during S phase, and only one of the two daughter cells resulting from cell division would contain this chromosome. In the next cellular generation, only one of four progeny cells would carry this chromosome, and so on, until the chromosome was effectively diluted out from the cell population.

(b)   Without a centromere, kinetochores would not form on the duplicated chromosome during mitosis, and the two chromatids would not segregate to different daughter cells. One daughter cell would end up with an extra copy of the chromosome, and the other cell would lack a copy.

(c)   With each round of DNA replication, the progeny DNA molecules would be slightly shortened at both ends, as is usually the case. However, at the end lacking a telomere for protection, the chromosome would presumably lose important genetic information in a relatively small number of cell generations. Even more dire consequences could result, especially if the chromosome in question were in a germline cell. The telomere lengthening enzyme, telomerase, would be unable to function at one end of the DNA, and eventually the DNA (and chromosome) would be completely degraded from that end.

19-8.   (a)   DP   (d)   DF   (g)   DP, DA, DF   (j)   None
        (b)   DF   (e)   None   (h)   DP, DA, DF
        (c)   DP   (f)   DA   (i)   DP, DA, DF

19-9.   (a)   Uracil arises from cytosine, hypoxanthine from adenine, and xanthine from guanine.

        (b)   Thymine does not contain an amino group and therefore cannot be deaminated.

        (c)   The excision process would not be able to distinguish between a base that has been generated by deamination and a naturally occurring base.

        (d)   The base 5-methylcytosine yields thymine upon deamination, and there is no mechanism to detect and excise thymine bases. Deamination of 5-methylcytosine will therefore go unrepaired, and every such deamination will convert a C into a T.

19-10.   (a)   Inconsistent. If chromosome movement were explained solely by microtubule disassembly at the spindle pole, the chromosomes would not move closer to the photobleached area. Instead, the photobleached area and the chromosomes would simultaneously move toward the spindle pole at the same rate.

        (b)   Consistent. As motor proteins located at the chromosomal kinetochore induce microtubule depolymerization, the chromosome is moved toward the spindle pole (and hence closer to the photobleached area) as the plus end of the microtubule is "chewed up."

        (c)   Consistent. However, while chromosome movement toward the photobleached area is consistent with movement driven by motor proteins that move along the microtubule and pull the chromosome along, it is also consistent with movement driven by motor proteins that act by depolymerizing ("chewing up") the plus end of the microtubule. The kinesin-like proteins located at the chromosomal kinetochore appear to act by the latter mechanism.

19-11.   (a)   C3 transferase blocks Rho, which is needed to stimulate assembly of the contractile ring. As a result, cytokinesis would be blocked. This is in fact exactly what happens when this experiment is performed.

(b) In nematode worm embryos lacking anillin, although Rho operates normally, inefficient assembly of an actomyosin network in the contractile ring would lead to failures in cytokinesis, which requires nonmuscle myosin II contraction of actin filaments in the contractile ring.

19-12. (a) G1 has the 2C amount of DNA; G2 has the 4C amount.

(b) DNA is being synthesized during S; little or no synthesis occurs during G1.

(c) Chromosomes are in an extended form during G2, but in a condensed form during most of M phase. A cell near the end of M phase may have decondensed chromosomes, but it will be distinguished by having two nuclei not yet completely partitioned into daughter cells.

(d) Chromosomes are in an extended form during G1, but in a condensed form during most of M phase. See also answer in part c.

19-13. (a) T (c) F (e) T

(b) F (d) NP

19-14. (a) Mitotic Cdk-cyclin is normally activated at the end of G2, when it catalyzes the phosphorylation of condensins and lamins. Phosphorylation of these proteins in turn contributes to chromosome condensation and nuclear envelope breakdown. It is therefore possible to trigger chromosome condensation and nuclear envelope breakdown prematurely by experimentally introducing mitotic Cdk-cyclin into cells that have just emerged from S phase.

(b) The ability of an indestructible form of cyclin to block the exit from mitosis shows that destruction of mitotic cyclin—and hence the inactivation of mitotic Cdk—are required before cells can complete mitosis and begin a new cell cycle.

(c) Breakdown of the nuclear envelope at the beginning of mitosis is triggered by the phosphorylation of lamins by mitotic Cdk-cyclin. The discovery that cells deficient in protein phosphatase activity have difficulty reconstructing the nuclear envelope suggests that proteins phosphorylated by mitotic Cdk-cyclin at the beginning of prophase must be dephosphorylated again before cells can finish mitosis.

19-15. (a) Since pifithrin-α reversibly blocks p53-dependent transcriptional activation, one might expect two effects: (1) p21 would not be expressed, leading to loss of cell cycle arrest, and (2) Puma would not be expressed, leading to *more* efficient inhibition of mitosis. In such cells, apoptosis would be less likely to occur, and they would be more likely to divide.

(b) TRAIL would activate pathways similar to TNF-α, a known stimulator of apoptosis.

(c) Caspase-3 is a key activator of the apoptosis pathway, so inhibiting it would suppress cell death.

CHAPTER

20

Sexual Reproduction, Meiosis, and Genetic Recombination

20-1. (a)   S

    (b)   A

    (c)   B. However, in sexual reproduction, a mutation in one of a homologous pair of chromosomes will likely be inherited by only some of the offspring (50%, on average).

    (d)   B. However, many more such offspring will result from sexual reproduction, where their occurrence is not dependent on new mutations, which are rare.

    (e)   B

20-2. (a) $2n = 4$

    (b)   Correct order: F, E, D, B, A, C.

    A: Metaphase II

    B: Telophase I (and cytokinesis)

    C: Anaphase II (and cytokinesis)

    D: Metaphase I

    E: Late prophase I (diplotene)

    F: Early prophase I (leptotene)

    (c)   Between metaphase I (D) and telophase I (B).

    (d)   Between leptotene (F) and diplotene (E) of prophase I.

20-3. (a)   Homologous chromosomes pair at meiotic metaphase I, but not at mitotic metaphase.

    (b)   Meiotic metaphase II has only one half as many chromosomes as does mitotic metaphase.

    (c)   Compared to metaphase I, metaphase II has only one half as many chromosomes and no paired bivalents.

(d)   Mitotic telophase yields two diploid daughter nuclei; meiotic telophase II yields four haploid daughter nuclei.

(e)   Chromosomes of each bivalent have begun to separate and chiasmata have become visible by diplotene, but not by pachytene.

20-4.   (a)   Yes. Each centromere is duplicated and one of each is passed to each of the daughter cells resulting from mitosis. All somatic cells arise from the zygote by mitosis, and the products of mitosis are always equal.

(b)   No. Centromeres separate at the first meiotic division, and independent assortment will randomize the segregation of maternal and paternal centromeres. Each first-division daughter cell will receive a maternal or a paternal centromere for each homologous pair.

20-5.   (a)   2X   (c)   2X   (e)   4X   (g)   2X

(b)   ½ X   (d)   ½ X   (f)   2X

20-6.   If the bivalent containing the two members of the chromosome 13 pair failed to separate at anaphase I of meiosis, this nondisjunction event would lead to the formation of some gametes containing two copies of chromosome 13 (as well as some gametes containing no copies of chromosome 13). If a gamete containing two copies of chromosome 13 were to fertilize a normal gamete (containing one copy of chromosome 13), the resulting zygote would have an extra copy of chromosome 13 (i.e., it would contain three copies of chromosome 13 instead of the normal two).

In a similar fashion, nondisjunction of the XY chromosome pair during anaphase I of male gamete formation, or nondisjunction of the XX chromosome pair during anaphase I of female gamete formation, could lead to the production of gametes containing neither an X chromosome nor a Y chromosome. If a sperm containing neither an X chromosome nor a Y chromosome were to fertilize a normal egg (which contains a single X chromosome), the result would be a zygote with only a single X chromosome. Similarly, if an egg containing no X chromosome were fertilized by a normal sperm containing an X chromosome, the result would be a zygote with only a single X chromosome.

20-7.   (a)   The four possible genotypes of the offspring, in their respective ratios, are 1 *YY*, 2 *Yy*, and 1 *yy*. Because all *YY* and *Yy* offspring will be yellow and only *yy* offspring will be green, offspring with yellow and green seeds will appear in the ratio 3:1.

(b)   Both parents are heterozygous for both factors, so the number of gametes is 2^2, or 4. A 4×4 matrix has 16 outcomes for offspring, although not all of the 16 are different. In general, the number of different gametes is 2^n, where n is the number of heterozygous allelic pairs.

(c)   The Punnett square assumes that the frequency of occurrence of a given genotype among the boxes represents the frequency of occurrence of that genotype among the progeny of the genetic cross represented by the Punnett square. This assumption is only correct if all possible combinations are equally likely, which in turn depends on an independent assortment of the alleles for seed color and seed shape.

(d)   The completed matrix will have 16 genotypes, nine of which are different from one another. In their respective ratios, these are 1 *YYRR*, 2 *YYRr*, 1 *YYrr*, 2 *YyRR*, 4 *YrRr*, 2 *Yyrr*, 1 *yyRR*, 2 *yyRr*, and 1 *yyrr*.

(e)   The phenotypes are yellow and round, yellow and wrinkled, green and round, and green and wrinkled. These occur in the ratio 9:3:3:1.

20-8.   (a)   The four genes are arranged as follows:

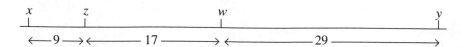

(b)   According to the map, the distance between genes *z* and *y* should be 17 + 29 = 46 map units, yet the data give a value of only 44% recombinants between these two genes. The observed value is slightly lower than expected because two recombination events (crossovers) can occasionally occur between the two genes, thereby reestablishing the parental combination. The farther apart two genes are located, the more likely that two crossover events will occur between them.

Also note that genes *x* and *y* are separated by 9 + 17 + 29 = 55 map units, yet the recombination frequency between them is only 50%. In fact, the percentage of recombinants can never exceed 50% because for genes that are assorting completely independently, an equal number of recombinants and parental combinations would be expected (i.e., a 50% recombination frequency). This will be observed if two genes are either on different chromosomes, or are located so far apart on the same chromosome that at least one crossover event is almost certain to occur between them. Because recombination frequencies cannot exceed 50%, genetic mapping is most accurate when recombination frequencies are measured on genes located relatively close to one another.

20-9.   (a)   RecBCD binds to double-strand breaks in chromosomal DNA and its helicase activity then unwinds the DNA double helix, thereby creating single-stranded loops. The nuclease activity of RecBCD cleaves one of the looped DNA strands, creating a free, single-stranded DNA end. RecA then catalyzes a "strand invasion" reaction in which this free, single-stranded DNA end invades and displaces one of the two strands of an homologous DNA double helix (step 2 in Figure 20-24 on p. 629 of the textbook).

(b)   While unwinding the DNA double helix with its helicase activity, the RecBCD protein moves along the DNA until it encounters a CHI site. It then stops at the CHI site and cleaves one of the DNA strands, thereby setting the stage for the RecA-catalyzed strand invasion step. Hence, recombination preferentially occurs at CHI sites in cells containing RecBCD.

20-10.   (a)   If recombination were a frequent occurrence, then the cloned DNA inserted into the plasmid would be subject to recombination, resulting in undesirable alterations in the sequence of the clone.

(b)   Restriction endonucleases are designed to remove viral DNA insertions into the bacterial genome as a defense mechanism. By mutating the genes in the bacterial genome that encode restriction endonucleases, it is much less likely for the foreign-cloned DNA to be excised.

CHAPTER

21

Gene Expression I: The Genetic Code and Transcription

21-1. (a) Because six and nine are multiples of three, there would be no effect on the reading frame. The only difference would be how many codons are removed or added (and hence how many amino acids might be missing or added).

(b) If the code involved sextuplets, then addition or deletion of three or nine bases would cause a frameshift, leading to widespread differences in amino acid sequence in the resulting polypeptide. In contrast, addition or deletion of six bases would not cause a frameshift.

21-2. (a) 5'-AAA(or G) AGUCCAUCACUUAAUGCN-3' (where N is any nucleotide).

(b) 5'-AAA(or G) GUCCAUCACUUAAUGGCN-3'.

(c) In terms of the mRNA, we know that an A was deleted and a G was inserted. In terms of the DNA template strand, these changes would correspond to a deleted T and an inserted C. However, we don't know *exactly* where along the DNA strand these changes occurred. Speaking again in terms of the mRNA, if the Lys codon at the 5' end of the original sequence was AAA, making the first four nucleotides AAAA, any of these A nucleotides could have been deleted. On the other hand, if the Lys codon was AAG, the deleted A must have been the next one in the sequence. The new G could have been inserted on either side of the G that is the third nucleotide from the 3' end.

21-3. (a) Not OK (d) Not OK (g) Not OK
     (b) OK   (e) OK   (h) OK
     (c) Not OK   (f) OK   (i) OK

21-4. Glu → Val   GAA → GUA
                or GAG → GUG

     His → Tyr   CAU → UAU
                or CAC → UAC

     Asn → Lys   AAU → AAA
                or AAC → AAG

     Glu → Gly   GAA → GGA
                or GAG → GGG

Ile → Val	AUU → GUU
	or AUC → GUC
	or AUA → GUA
Tyr → Cys	UAU → UGU
	or UAC → UGC
Gly → Arg	GGA → CGA
	or GGG → CGG
Lys → Stop	AAA → UAA
	or AAG → UAG

21-5. (a) These data suggest that the promoter for the 5S rRNA gene is located between nucleotides 47 and 83, because deletions that include nucleotides within this stretch prevent RNA polymerase III from transcribing the gene. Note that this type of promoter lies *within* the gene—that is, it is located downstream from the startpoint.

(b) Promoters for RNA polymerase I typically lie in the region between –45 and +20 (see Figure 21-11a on p. 662 of the textbook). Therefore, if a gene transcribed by RNA polymerase II had deletions in the same locations as the first four entries listed in the table (–45 through –1, or +1 through +47, or +10 through +47, or +10 through +63), transcription would not be expected to take place because some of the nucleotides between –45 and +20 would be missing. Deletions corresponding to the last two entries in the table (+80 through +123 and +83 through +123) would not be expected to affect promotion.

(c) Promoters for RNA polymerase II are of two general types. (1) TATA-driven promoters contain an Inr sequence and a TATA box, with or without an associated BRE immediately upstream from the TATA box. Therefore, if a gene transcribed by RNA polymerase II had deletions in the same location as the first two entries in the table (–45 through –1 or +1 through +47), transcription would not be expected to occur because either the TATA box or the Inr sequence would be disrupted. Deletions corresponding to the last four entries in the table would not be expected to affect promotion because they are all downstream of BRE, the TATA box, and Inr. (2) DPE-driven promoters contain Inr sequences near the startpoint and DPE sequences about 30 nucleotides downstream from Inr. Therefore, deletions corresponding to the first two entries in the table (–45 through –1 or +1 through +47) would both fail to be transcribed because the Inr sequence would be disrupted. In addition, deletions corresponding to the second, third, and fourth entries in the table (+1 through +47, +10 through +47, and +10 through +63) would fail to be transcribed because the DPE sequence would be disrupted. Deletions corresponding to the last two in the table would not be expected to affect promotion because they are all downstream from Inr and DPE.

21-6.

(a)	B, I	(g)	None
(b)	B, I, II, III	(h)	B, I, II, III
(c)	I, II, III	(i)	II
(d)	B	(j)	B, I, II (Only RNA polymerase III promoters may lie
(e)	B, I, III		entirely downstream of the transcriptional startpoint.)
(f)	I, II, III	(k)	B

21-7. (a)   rRNA: Cleavage of large precursor; degradation of transcribed spacers.

tRNA: Removal of leader sequence at 5′ end; addition of CCA sequence at 3′ end (if not already present); methylation of bases; removal of introns.

mRNA: Elimination of introns; addition of poly(A) tail on 3′ end; capping of 5′ end.

(b)   rRNA: Generation of several molecules from one large precursor.

tRNA: Addition of CCA sequence at 3′ end if not already present.

mRNA: Capping of 5′ end; addition of poly(A) tail.

(c)   rRNA: Generation of several molecules from one large precursor; degradation of transcribed spacers.

tRNA: Addition of CCA sequence if necessary; methylation of bases; cleavage from a larger precursor.

mRNA: None in common.

21-8. (a)   R; snRNA are the small (nuclear) RNA molecules that are incorporated into the spliceosome. In addition to a structural role in spliceosome assembly, snRNA molecules recognize (by base pairing) the splice sites on the primary RNA transcript and probably play a major role in catalyzing the splicing reaction.

(b)   PR; spliceosomes, the large molecular complexes where splicing occurs, consist of snRNPs (see part c) and additional proteins.

(c)   PR; a variety of snRNPs, each consisting of one or two molecules of snRNA plus proteins, assemble with other proteins to form a spliceosome.

(d)   R; splice sites are the specific nucleotide sequences in the primary RNA transcript at the junctions between exons and introns. The spliceosome recognizes these sequences and cuts and splices at particular points within them.

(e)   R; when splicing occurs, the RNA of each intron leaves the transcript in the form of a lariat, which is subsequently degraded.

21-9. (a)   Rifamycin interferes with initiation because it prevents formation of the first phosphodiester bond.

(b)   Actinomycin D interrupts elongation because it blocks the DNA.

(c)   Actinomycin D, because rifamycin is specific for bacterial polymerase.

(d)   Rifamycin, because it inhibits initiation but does not block elongation.

(e)   Actinomycin D inhibits transcription, which will prevent the zygotes from making any new mRNAs. The reason that the early zygotes can develop for a significant period of time is because mRNAs are stored in the unfertilized egg. These mRNAs are placed into the egg during oogenesis, and so they are called "maternal" mRNAs, because they are derived from the mother. When the eggs are fertilized, these mRNAs are used to translate proteins needed during early development. Eventually, however, these maternal mRNAs undergo turnover, and are degraded. Because the zygote cannot make new mRNAs needed for continued development, the embryos arrest.

21-10. (a) Codons: UUU, UUC, UCU, CUU, CCU, CUC, UCC, CCC.

Amino acids: phenylalanine (UUU, UUC), serine (UCU, UCC), leucine (CUU, CUC), proline (CCU, CCC).

(b) Incubation A: 8 of each.

Incubation B: 27 UUU; 9 each of UUC, UCU, and CUU; 3 each of CCU, CUC, and UCC; and 1 CCC.

(c) Incubation A: 16 of each.

Incubation B: 36 phenylalanine; 12 each of serine and leucine; 4 proline.

(d) It is possible to distinguish between classes of codons that differ in their base composition and to determine the base compositions of the codons that code for various amino acids.

(e) Yes, the data of parts b and c for incubation B should allow this to be deduced.

(f) No.

(g) Synthesize the relevant codons and test them in the rRNA-binding assay of Nirenberg.

21-11. (a) At least one intron must be present in gene X. This intron contains a restriction site for *Hae*III, explaining why digestion with *Hae*III yields an extra fragment when the gene rather than the cDNA is cleaved with *Hae*III. It is possible that more than one intron is present in the gene, because the additional introns might not contain restriction sites for *Hae*III, and hence their presence could not be detected by digesting the DNA with *Hae*III.

(b) Although two extra fragments are produced when the gene rather than the cDNA is cleaved with *Hae*III, this does not necessarily mean that two introns are present in gene Y because the two additional *Hae*III sites could both be located in the same intron. Hence, you can only conclude that *at least* one intron is present.

(c) Because no extra fragments are produced when the gene rather than the cDNA is cleaved with *Hae*III, this result does not provide any evidence for the existence of an intron within gene Z. However, it is possible that one or more introns are present in the gene, because these introns might not contain restriction sites for *Hae*III, and hence their presence could not be detected by digesting the DNA with *Hae*III.

21-12. Bacteria do not use splicing to edit RNA after its transcription. The inserted genomic DNA ("gene") from the eukaryotic liver cell contains intronic and exonic sequences. When the bacterium produces mRNA from this sequence, it will not be spliced, and as a result, when the RNA is translated it will contain incorrect amino acids (and/or an early stop).

CHAPTER

22

Gene Expression II: Protein Synthesis and Sorting

22-1. (a) 5' UUAAU<u>AUG</u>UGCUACUUCGAACACUGUCCCAAAGG<u>UUAG</u>U<u>AA</u>UU 3'

3' AAUUAUACACGAUGAAGCUUGUGACAGGGUUUCCAAUCAUUAA 5'

(b) Only the first of the two RNA sequences has an initiation codon (AUG, under-scored above) in the correct (5' → 3') direction, so it must be the messenger. Notice also that the same RNA sequence has two stop codons in the correct reading frame near the 3' end (UAG, UAA, also underscored).

(c) The amino acid sequence of vasopressin is, therefore,

```
AUG UGC UAC UUC GAA CAC UGU CCC AAA GGU  UAG  UAA

Met-Cys-Tyr-Phe-Glu-His-Cys-Pro-Lys-Gly-(stop)

(0)  1   2   3   4   5   6   7   8   9
```

(d) The methionine at the N-terminal end is apparently cleaved to generate the mature polypeptide.

(e) Oxytocin differs from vasopressin in positions 3 (Ile for Phe), 5 (Asp for His), and 8 (Leu for Lys). The first two amino acid changes can be accomplished by one-base changes in the mRNA (AUC instead of UUC, GAC instead of CAC), but the third requires a two-base change (UUA or CUA instead of AAA). The most conservative DNA sequence would therefore be

3' AATTATACACGATGTAGCTTCTGACAGGGAATCCAATCATTAA 5'

The similarity between the coding sequences for the two hormones suggests an evolutionary relationship between them. Either one evolved from the other, or both evolved from a common ancestral sequence.

22-2. (a) Arg: AGA
Ile: AUA
Lys: AAA
Stop: UAA
Thr: ACA
Ser: UCA

(b)	UAU (or C)	Tyr	1st position
	UCA	Ser	2nd position
	UUA	Leu	2nd position
	CAA	Gln	3rd position
	AAA	Lys	3rd position
	GAA	Glu	3rd position

22-3.   His (codon CAU fulfills these criteria).

22-4.   In the initiation of eukaryotic translation, the first few binding steps occur in a sequence different from that in bacterial initiation. In the bacterial case, initiation factors and GTP attach to a small ribosomal subunit before the initiator tRNA (carrying fMet) arrives. In the eukaryotic case, an initiation factor binds to GTP and to the initiator tRNA (carrying Met) before any of them bind to the small ribosomal subunit. The way in which the small ribosomal subunit arrives at the mRNA start codon also differs. In bacteria, the ribosome binds to a ribosome-binding site (Shine-Dalgarno sequence) in the mRNA, which is located near the start codon. In eukaryotes, the small ribosomal subunit–tRNA complex initially binds to the mRNA at its 5′ cap and then moves down the mRNA until it reaches the start codon, where the tRNA anticodon base-pairs with the start codon. Only then does the large ribosomal subunit join the complex.

22-5.   (a)   Ba   (c)   Bo   (e)   Bo   (g)   Bo   (i)   Bo
       (b)   Bo   (d)   N   (f)   N   (h)   Bo

22-6.   (a)   Because it resembles the end of an aminoacyl tRNA molecule, puromycin can bind to the ribosome and form a peptide bond with the carboxyl group of the growing polypeptide chain. The product, peptidyl puromycin, then dissociates from the ribosome.

        (b)   To the carboxyl end, because it has an amino group.

        (c)   To the A site, because it resembles an aminoacyl tRNA.

        (d)   Equally effective, because it interferes with the polypeptide elongation stage of protein synthesis, which is similar in bacterial and eukaryotic cells.

22-7.   (a)   Rambomycin inhibits translocation of peptidyl tRNA from the A site to the P site, with concomitant movement of the mRNA. Inhibition of any earlier step would prevent formation of the first peptide bond.

        (b)   Both products will remain as peptidyl tRNA because there is no stop signal present to activate a release factor.

22-8.   A mutant tRNA with a *four*-nucleotide anticodon could suppress a frameshift mutation consisting of a one-nucleotide insertion nearby. A mutant tRNA with a *two*-nucleotide anticodon could suppress a frameshift mutation consisting of a one-nucleotide deletion.

22-9.   (a)   BiP contains a KDEL sequence, causing it to be retained in the ER.

        (b)   If BiP lost the ability to bind to hydrophobic amino acids, it could no longer bind to unfolded and misfolded polypeptides. The exposed hydrophobic regions of these unfolded and misfolded polypeptides would cause them to aggregate with each other, forming insoluble deposits that can trigger disruptions in cell function and may even lead to cell death.

(c)   Although it is possible that polypeptide X might be stable in the absence of polypeptide Y, the different polypeptides that make up a multisubunit protein must often interact with each other in order to fold into their proper three-dimensional conformation. Thus, in the absence of polypeptide Y, polypeptide X might remain bound to BiP because it cannot fold properly in the absence of polypeptide Y. Polypeptides that repeatedly fail to fold properly are eventually shuttled back across the ER membrane to the cytoplasm, where they are degraded.

22-10. (a)   In the absence of SRP and ER membranes, a prolactin *precursor* molecule is synthesized. This precursor, called preprolactin, consists of the normal prolactin sequence (199 amino acids) plus a 28 amino acid ER signal sequence at the N-terminus.

(b)   SRP binds to the ER signal sequence of the newly forming polypeptide chain and halts protein synthesis, thereby preventing the chain from growing beyond 70 amino acids. Inside cells, this blockage is normally maintained until the SRP binds the ribosome-mRNA complex to the ER. Without this blockage, a complete preprolactin chain might be produced by cells and released into the cytosol, rather than being transported into the ER lumen for secretion from the cell.

(c)   In the presence of ER membrane vesicles, the SRP binds the mRNA-ribosome complex to the ER, where signal peptidase cleaves the ER signal sequence from the newly forming prolactin chain as it moves across the ER membrane. Removal of the ER signal sequence reduces the length of the final polypeptide from 227 amino acids to 199 amino acids. The final polypeptide chain is released into the lumen of the ER vesicles.

22-11. (a)   B   (e)   Nu
         (b)   M   (f)   M
         (c)   M   (g)   B
         (d)   M   (h)   Nu

CHAPTER

23

The Regulation of Gene Expression

23-1. Activity of the z gene:

	Presence of Lactose	Absence of Lactose	Explanation
(a)	+	−	Wild-type; inducible system
(b)	−	−	Superrepressor cannot recognize inducer
(c)	+	+	Operator cannot bind repressor
(d)	+	+	Repressor not made; system always "on"
(e)	+	+	Operator cannot bind repressor
(f)	−	−	Promoter cannot bind RNA polymerase
(g)	−	−	Glucose reduces CRP binding

23-2. (a) Genes A and B code for enzymes involved in the synthesis of ethanol, C is the operator, and D is the gene encoding the repressor.

(b) (i) No synthesis of enzyme A, inducible for enzyme B; cannot produce ethanol.

(ii) Inducible for both enzymes; wild-type phenotype.

(iii) Inducible for both enzymes; wild-type phenotype.

(iv) Inducible for enzyme A, constitutive for enzyme B; wild-type phenotype.

23-3. (a) C (c) X (e) C (g) C (i) I
(b) C (d) I (f) I (h) C

23-4. (a) Its mode of action is to terminate transcription prematurely to varying extents, based on the concentration of tryptophan in the cell.

(b) Signals termination, causing dissociation of the RNA polymerase from the template.

(c) The transcriptional activity of an RNA polymerase molecule is influenced by changes in mRNA structure induced by a ribosome that is translating that mRNA.

(d)   It ensures that the ribosome does indeed "tailgate" the RNA polymerase, allowing the necessary proximity of the ribosome to the RNA polymerase molecule.

(e)   Has amino acid X present at more than one position.

(f)   The model requires sensitivity of transcription to aminoacyl tRNA availability and therefore to the presence of a specific amino acid in the leader peptide.

(g)   The close coupling of transcription and translation on which the model depends is not possible in eukaryotes because of the nuclear envelope.

23-5.   (a)   B, G   (c)   N   (e)   L   (g)   L
         (b)   L   (d)   G   (f)   N   (h)   N

23-6.   (a)   True. Enhancers have been shown to increase rates of transcription initiation even when located up to tens of thousands of base pairs from the promoter that they affect.

         (b)   True. In a few cases, enhancers located within genes have been identified and observed to increase transcription levels. For example, the introns of some immunoglobulin genes contain enhancers.

         (c)   False. Regardless of an enhancer's position, it cannot increase the rate of transcription of a gene unless the gene has a core promoter. The core promoter region provides the binding site for TFIID and RNA polymerase.

         (d)   True. Enhancers appear to act by forming a loop of DNA as the activators bound to the enhancer bind, via different domains, to coactivators that are subunits of TFIID, which in turn binds to the promoter.

         (e)   False. Enhancers have not been shown to play a role in DNA splicing.

23-7.   (a) $(500)(13,000)/(2.7 \times 10^9) = 0.0024 = \mathbf{0.24\%}$.

         (b) $(2.7 \times 10^9) + (5 \times 10^5)(13,000) = \mathbf{9.2 \times 10^9}$ **base pairs** per amplified haploid genome.

                Ribosomal genes represent $6.5/9.2 \times 100\%$, or **70%,** of the amplified genome!

         (c)   Because amplification is 1000-fold, the unamplified genome would presumably require 1000 times longer to synthesize the required number of ribosomes. Therefore, 2000 months, or about **167 years,** would be needed (which would make female frogs *very* old mothers!).

         (d)   When the desired gene product is an RNA, all the molecules the cell needs must be transcribed directly from the DNA. When the desired product is a protein, its mRNA can be translated repeatedly, so there is an additional stage of amplification at the mRNA level.

23-8.   (a)   Testosterone and hydrocortisone bind to different hormone receptors present in liver cells. The binding of testosterone allows its receptor to bind to testosterone response elements in DNA, whereas the binding of hydrocortisone allows its receptor to bind to glucocorticoid response elements in DNA. The gene coding for α2-microglobulin is associated with a testosterone response element, whereas the gene coding for tyrosine aminotransferase is associated with a glucocorticoid response element. As a result, testosterone and hydrocortisone activate the transcription of the genes coding for α2-microglobulin and tyrosine aminotransferase, respectively.

(b)   Steroid hormones such as testosterone and hydrocortisone bind to receptors that can activate gene transcription, leading to the production of mRNAs that are then translated into new polypeptide chains. Therefore, inhibitors of either mRNA synthesis (e.g., α-amanitin) or protein synthesis (e.g., puromycin) would be expected to block the ability of these two hormones to stimulate the production of α2-microglobulin and tyrosine aminotransferase, respectively.

(c)   The zinc-finger domain is responsible for the ability of hormone receptors to recognize and bind to the specific DNA sequences that make up their corresponding DNA response elements. Therefore, switching the zinc-finger domain between the testosterone and glucocorticoid receptors would be expected to switch the DNA-binding specificities of the two receptors. As a result, the testosterone receptor would bind to glucocorticoid response elements, and the hydrocortisone receptor would bind to testosterone response elements. The binding of testosterone to its receptor would now be expected to increase the production of tyrosine aminotransferase, and the binding of hydrocortisone to its receptor would be expected to increase the production of α2-microglobulin.

(d)   Testosterone would increase the production of both α2-microglobulin and tyrosine aminotransferase, whereas hydrocortisone would increase the production of neither of these proteins.

23-9.   (a)   False. Homeotic genes play key roles in early *Drosophila* development by directing synthesis of transcription factors that coordinate the development of fundamental characteristics such as body shape and appendage location. Furthermore, homeotic proteins contain helix-turn-helix domains, not zinc fingers.

(b)   False. Mutations in homeotic genes are not necessarily fatal for *Drosophila*, but they often lead to very dramatic abnormalities, such as legs sprouting from the fly's head.

(c)   True. Homeotic proteins form a family of widely active transcription factors, each of which may bind to and control expression of hundreds of other genes in a *Drosophila* embryo.

(d)   False. They regulate transcription of mRNA.

23-10.   Mammals have four, partially degenerate, sets of *Hox* genes, whereas flies only have a single set. Because in some cases more than one of the mammalian *Hox* clusters encodes a similar protein, mutations in some of the mammalian *Hox* genes have mild phenotypes. This is an example of functional redundancy, that is, a situation in which loss of one protein is compensated for by other proteins.

23-11. (a)   One possibility is that the missing stretch of amino acids contains the lysine to which ubiquitin is normally attached when mitotic cyclin is being targeted for destruction. A second possibility is that the missing segment contains a degron whose detection by an appropriate recognition protein normally targets mitotic cyclin for ubiquitylation. Finally, the removal of a stretch of amino acids could change the conformation of mitotic cyclin so that it can no longer serve as a substrate for the ubiquitylating enzyme complex.

(b)   The amino acid sequence of the ubiquitylation site of normal mitotic cyclin could be analyzed to determine whether the lysine residue that serves as an attachment site for ubiquitin is missing in the mutant cyclin. Alternatively, recombination DNA techniques could be used to insert the missing stretch of nine amino acids into other proteins to see whether it can serve as a degron that targets these proteins for degradation. If the preceding kinds of experiments fail to provide evidence that the deleted stretch of amino acids serves as either a ubiquitylation site or a degron, then it is possible that that amino acid deletion is simply exerting its effects by altering protein conformation, although this hypothesis would be difficult to prove directly.

(c)   A mutation in a recognition protein that specifically binds to a degron present in mitotic cyclin would explain such observations.

23-12. (a)   The total mRNA population of one tissue could be radioactively labeled and hybridized against total mouse DNA (or, for greater sensitivity, against the nonrepeated DNA component of the mouse genome) in the presence and absence of mRNA from the other tissue. The extent to which radioactive RNA from one tissue could be "competed out" by nonradioactive species from the other tissue measures the fraction of common sequences. Alternatively, DNA microarrays could be used to compare the mRNAs produced in the two tissues.

(b)   Selective processing of nuclear transcripts and selective translocation of RNA to the cytoplasm could also explain the data. To distinguish between these possibilities, nuclear RNA preparations from the two tissues could be tested for the presence or absence of common sequences by the same hybridization techniques used with the cytoplasmic RNA.

(c)   Translational control would seem likely.

23-13.   As part of the normal process of female gametogenesis, sea urchin eggs produce and store large amounts of mRNA. After an egg is fertilized by a sperm cell, these mRNAs are translated into the proteins that are required during the early stages of embryonic development. Because these mRNAs are already present, inhibitors of RNA synthesis have relatively little effect on either protein synthesis or early development. Later in development, genes coding for proteins required during the later stages of embryonic development begin to be transcribed into mRNAs that are then translated into polypeptides. At this point, blocking transcription with inhibitors of RNA synthesis will prevent these essential proteins from being produced.

CHAPTER

24

Cancer Cells

24-1.   (1) If each sample of cells is injected into nude mice, you would expect only the cancer cells to produce tumors. (2) If the cells are grown in culture, only the cancer cells would be expected to be anchorage independent—that is, capable of growing well in liquid suspension or in soft agar. (3) The cancer cells would be expected to grow to higher population densities in culture because of the loss of density-dependent inhibition of growth. (4) Normal cells would be expected to divide no more than 50–60 times in culture, whereas cancer cells do not exhibit such a limit because they have mechanisms for replenishing their telomeres.

24-2.   Evidence that angiogenesis is required for tumors to grow beyond a tiny clump of cells is as follows: (1) Cancer cells injected into an isolated rabbit thyroid gland kept alive with a nutrient solution fail to link up to the organ's blood vessels, and the tumor stops growing when it reaches a diameter of roughly 1–2 mm. When injected into live animals, these same tumors became infiltrated with blood vessels and grow to an enormous size. (2) Cancer cells placed in the anterior chamber of a rabbit's eye, where there are no blood vessels, remain alive but stop growing before the tumor reaches 1 mm in diameter. In contrast, cancer cells placed directly on the iris become infiltrated with blood vessels and the tumors grow to thousands of times their original mass.

Evidence supporting the idea that cancer cells secrete molecules that stimulate angiogenesis is as follows: (1) When cancer cells are placed in a chamber surrounded by a filter possessing tiny pores that cells cannot pass through, and the chamber is then implanted into animals, new capillaries proliferate in the surrounding host tissue. Since the cancer cells cannot pass through the filter, the most straightforward interpretation is that the cells produce molecules that diffuse through the tiny pores in the filter and activate angiogenesis in the surrounding normal tissue. (2) Cancer cells produce and secrete angiogenesis-activating proteins called *vascular endothelial growth factor (VEGF)* and *fibroblast growth factor (FGF)*, which bind to receptor proteins on the surface of endothelial cells. The activated endothelial cells then organize into hollow tubes that develop into new blood vessels.

24-3.   The three main steps are (1) cancer cells invade surrounding tissues and penetrate through the walls of lymphatic and blood vessels, thereby gaining access to the bloodstream; (2) the cancer cells are transported by the circulatory system throughout the body; and (3) cancer cells leave the bloodstream and enter particular organs, where they establish new metastatic tumors. The first step is facilitated by decreased cell–cell adhesion, increased motility, and secretion of proteases that degrade the extracellular matrix and basal lamina. During the second step, only a tiny fraction of the cancer cells that enter the bloodstream survive the trip and establish successful metastases. To some extent, the limited survival of cancer cells during this stage may be influenced by the ability of the immune system to attack and destroy cancer cells. During the third step, the sites to which cancer cells tend to metastasize are determined by the location of the first capillary bed encountered as well as organ-specific conditions that influence cancer cell growth.

24-4.   AAF is a "precarcinogen" that needs to be metabolically activated before it can cause cancer, which explains why it does not cause cancer when isolated cells from either rats or guinea pigs are exposed to it. Rats, but not guinea pigs, contain a liver enzyme that catalyzes the metabolic activation of AAF, which explains why it causes cancer when injected into rats but not guinea pigs. Analyzing human liver cells to see if they contain the activating enzyme would indicate whether AAF is likely to be carcinogenic in humans.

24-5.   (a)   Proto-oncogenes and tumor suppressor genes are both found in normal cells. Oncogenes are genes whose presence can cause cancer, and so are not found in normal cells.

(b)   Some proto-oncogenes code for normal growth factors (e.g., the proto-oncogene that codes for PDGF). Although oncogenes often code for abnormal versions of proto-oncogene proteins, they can also code for excessive quantities of a normal protein encoded by a proto-oncogene. Hence, an oncogene could code for excessive amounts of a normal growth factor.

(c)   Oncogenes can obviously be found in cancer cells. However, cancer cells also contain many normal proto-oncogenes and tumor suppressor genes. Only a small number of the dozens of proto-oncogenes and tumor suppressor genes in a normal cell need to be mutated to trigger the development of cancer.

(d)   Oncogenes are the only genes that are uniquely present in cancer cells.

(e)   Oncogenes are genes whose presence can cause cancer.

(f)   Tumor suppressor genes are genes whose absence can cause cancer.

(g)   Tumor suppressor genes and proto-oncogenes are present in both normal cells and cancer cells. (Their presence in cancer cells is explained by the fact that only a small number of the dozens of proto-oncogenes and tumor suppressor genes present in a normal cell need to be mutated to trigger the development of cancer.)

24-6.   Pyrimidine dimers caused by exposure to sunlight can create mutations that lead to cancer. One of the main mechanisms for repairing such defects is excision repair, a process that recognizes distortions in the DNA double helix, excises the damaged region, and fills in the resulting gap with the correct sequence of nucleotides. Because individuals with xeroderma pigmentosum are usually unable to carry out excision repair, mutations accumulate and cancer eventually arises. The word "usually" is included in the previous sentence because inherited mutations in the gene coding for DNA polymerase η (eta) cause a variant form of xeroderma pigmentosum in which excision repair remains intact. DNA polymerase η is a special form of DNA polymerase that catalyzes the translesion synthesis of a new, error-free stretch of DNA across regions in which the template strand is damaged. So inherited defects in DNA polymerase η, like inherited defects in excision repair, hinder the ability to repair pyrimidine dimers.

24-7. A "false positive" rate of 2% means that if 100,000 people were randomly tested once per year using the FOBT test, the result would be 2000 false positives (2% of 100,000). But the average annual incidence of colon cancer in the United States is only 55 cases per 100,000. Therefore, the 2% false positive rate yields so many incorrect results (2000) compared to real cancer cases (55) as to make the procedure impractical for random testing. On the other hand, colon cancer rates increase dramatically as people get older, so if FOBT testing is restricted to older populations, the ratio of real cancer cases to false positives increases significantly, and the test becomes more useful.

24-8. Although antibodies directed against CD20 are toxic to normal lymphocytes, CD20 is not present on the precursor cells whose proliferation gives rise to these lymphocytes. As a result, proliferation of these precursor cells can replenish the normal lymphocytes inadvertently destroyed by antibodies directed against CD20.